D1221679

High Technology Small Firms

High Technology Small Firms

**Regional Development in Britain
and the United States**

Ray Oakey

St. Martin's Press, New York

338.762
0 11

Library of Congress Cataloging in Publication Data

Oakey, R. P. (Raymond P.)
     High Technology Small Firms.

     Bibliography =: P.
     Includes Index.
     1. Small business--United States--Case studies.
2. Small business--Great Britain--Case studies. 3. High
technology industries--United States--Case studies.
4. High technology industries--Great Britain--Case studies.
5. Technological innovations--Economic aspects--United
States--Case studies/. 6. Technological innovations--
Economic aspects--Great Britain--Case studies. I. Title.
HD2346.U5035 1984   338.7'6213817'0941   83-40705

ISBN 0-312-37239-6

Contents

List of tables

List of figures

Acknowledgements

This book could not have been produced without the help of several individuals to whom I must give thanks. Taking events chronologically, among many helpful colleagues in the Centre for Urban and Regional Development Studies, I must particularly thank Alfred Thwaites for initiating our work on technological change at Newcastle upon Tyne, for his collaboration on the earlier research projects mentioned in Chapter 2, and for his continuing overall support. Similarly, I also thank John Goddard for providing wide-ranging academic and logistical advice and assistance. During the execution of the survey work in Britain and the United States I was greatly indebted to my wife Dorothy, who organised a very congested appointments diary and performed various other secretarial duties.

With regard to the preparation of the text, I thank Roy Rothwell and David Storey for allowing me to use extracts from their work. Moreover, I must thank David Storey for detailed comments on earlier drafts of the book and Tony Champion for general comments. Any errors and omissions, however, remain my sole responsibility. Special thanks are due to Margaret Riley, who 'translated' illegible drafts into a final manuscript with remarkable efficiency and diligence.

Finally, I would like to acknowledge the support of the Social Science Research Council for funding the research on which this book is based.

1 Introduction

This book examines the innovation process in small independent high technology firms. Within this wider investigation, detailed consideration is given to the manner in which innovation in these small firms is affected by diverse local industrial environments in Britain and the United States. A detailed examination of the innovation environment of small high technology firms is timely, since the myths surrounding this type and size of industrial production have been largely accepted in the absence of detailed empirical evidence. 'Experts' in many parts of the Western industrialised economy have seized on the rapid, spontaneous growth apparent in the high technology small firms of Route 128 and Silicon Valley in the United States and proposed their replication as a panacea for the decline in industrial employment in depressed industrial areas. This new enthusiasm for high technology small firms has played a prominent role in the noted revitalisation of the Small Business Administration (SBA) in the United States (Thompson 1983), and in British government efforts aimed at small firm promotion (Storey 1983).

In many ways, however, the unassisted birth and growth of small firms in the now famous high technology industrial concentrations of the United States is an inappropriate model for the government assistance of small firms. Fundamentally, the difference in ethos between the 'free market' venture capital-based support produced in such areas and the 'drip feed' of government aid might initially seem at opposite ends of a spectrum of small firms' finance. Indeed, it might be argued that there is nothing in the growth experiences of high technology small firms in Silicon Valley or on Route 128 that might prompt modifications to grant-based aid offered by the SBA in other depressed areas of the United States, or to the various Department of Industry schemes in Britain.

Moreover, if firms in these established high technology industrial areas are examined closely to discover the reasons for this rapid growth, it will become clear that the local availability of capital is only a part, albeit an important part, of the total explanation of the prosperity which has been achieved. Thus since financial incentives are the main policy instrument of government small firm promotion agencies seeking to replicate Silicon Valley, it seems likely that current policies will provide only one of the resources necessary for the generation of concentrated high technology industrial growth. Indeed, as intimated above, even the capital-based incentives of government agencies differ markedly in spirit and function from the venture capital available in the high technology industrial areas.

Before national and local government agencies in both Britain and the United States devote large amounts of capital to the promotion of new Silicon Valleys and science parks in declining regions, there must be a clear understanding of *all* the conditions that create such areas. It is accepted that governments have the power to inject investment capital into a depressed region in an attempt to compensate for the 'free market' venture capital available in prosperous high technology areas. However, the creation of cumulative local advantages such as skilled white- and blue-collar labour, specialist material linkages, and technical information sources may require a more subtle long-term policy approach directed at building up a resource infrastructure over a number of years. The introduction of financial assistance in these circumstances must be based on an understanding of growth processes at work and on a commitment to a long-term development strategy for the whole region; success will not be achieved by piecemeal schemes directed at individual firms. Much of the enthusiasm for high technology industry stems from a rather superficial observation of the effects of such desirable growth rather than from a detailed appreciation of its causes.

However, despite these reservations, development agencies in both Britain and the United States are currently seeking new high technology production to diversify stagnant or declining local economies. This book will argue that such an objective *is* a worthwhile strategy, provided that it is not at the expense of traditional indigenous industries in decline. Clearly, small firm growth in expanding high technology sectors of industry has great potential for the absorption of a substantial proportion of the labour shed from irrevocably declining industries in the remaining decades of this century. However, an environment conducive to the formation and growth of high technology industrial production will not be achieved by merely renaming dilapidated industrial estates as science parks, or by the more vigorous extension of *existing* small firm aid. New policies must be conceived which are based upon rigorous investigations of the various contributory aspects of the total local economic advantages of high technology industrial agglomerations.

To meet the objectives of a comprehensive analysis of the impact of the local resource environment on high technology small firm innovation as a prerequisite of relevant policy, this book is divided into three unequal parts. The first part, comprising Chapters 2 to 4, considers relevant conceptual aspects of industrial location as it applies to high technology small firm formation and growth. Chapters 5 to 9 are based on empirical evidence. Chapter 5 relies predominantly on published statistical information from both Britain and the United States to provide a contextual picture of recent growth in the study instrument and electronic component industries in the survey regions. Subsequently, Chapters 6 to 9, based on survey data, examine local resource factors critical to innovation in high technology small firms; these are linkages, research and development (including sources of local technical information), labour, and innovation finance.

The survey data are derived from 174 British and American high technology firms in the San Francisco Bay area of California, Scotland, and the South-Eastern planning region of England. In some respects, Chapters 6 to 9 are autonomous in that they include their respective resource topics with conceptual introductions and relevant, specific conclusions. However, the book is held together by the recurrence of several theoretical principles. These accumulated principles, together with the survey results, are synthesised in the final chapter, which deals with the theoretical and policy implications of the book. In particular, the conclusion will address the issue of how the lessons learnt from the growth of small firms in high technology industrial concentrations might offer scope for new government policies to encourage high technology development in currently depressed regions.

2 Small firm employment, innovation and industrial growth

2.1 The international context

The recent acknowledgement by many Western politicians of the contribution of small firms to national industrial employment, and the missionary zeal that these new advocates have directed at policy, is somewhat disconcerting given the long-standing evidence on small firms and their substantial contribution to industrial employment. Certainly, the increased enthusiasm for small firm assistance in Britain derived little impetus from the earlier investigations of the Bolton Committee. For while its consideration of the role of small firms in the modern economy was comprehensive (Bolton 1971), the committee concluded that positive government discrimination in favour of small firms was not indicated and that small firm growth would revive 'naturally', due partly to the specialist market niches afforded by the demands of an increasingly affluent Western society. But, although the following evidence on the recent performance of small firms is frequently contradictory, there is little evidence that the decline of small firms in Britain has been arrested, far less reversed. Thus the British government's interest in small firm assistance is warranted. However, such interest mainly derives from a misguided belief that small firms are a panacea for national and regional unemployment, rather than from a more realistic conviction that small firm assistance is required in order to prevent the further contraction of this important scale of production.

Much of the confusion over the health of the small firm sector and its capacity for employment generation derived from a misinterpretation of the Birch report (Birch 1979), which aggregated data on American services and manufacturing employment growth. Many early media interpretations of these findings assumed that all the employment growth was in manufacturing, thus overstating the employment-generation potential of small *manufacturing* firms (Fothergill and Gudgin 1979). However, while the hard, published evidence of Table 2.1 on the contribution of small manufacturing firms to a range of national economies confirms their importance to *existing*

Table 2.1 Contribution of smaller enterprises to employment in manufacturing

Country	Year of data	% of total manufacturing employment in small firms	
		1-99 (%)	100-199 (%)
UK	1976	17.1	22.6
Denmark *†	1973	36.3	52.1
Ireland †	1968	33.0	49.8
Italy *	1971	n.a.	47.3
Netherlands	1973	36.0	47.8
Belgium	1970	33.2	43.4
Germany	1970	28.8	37.0
France	1976	25.2	34.2
Luxembourg	1973	19.1	28.2
Japan	1970	51.6	n.a.
USA	1972	24.8	n.a.

* Includes energy and water industries.
† Figures for these countries exclude employment in very small enterprises and understate the contribution of small enterprises.

Source: Commission of the European Communities (1980); Department of Industry (1976); Storey (1982).

employment, the aggregate short-term employment-generation *potential* of such firms, a motive for much recent political interest, must remain in serious doubt.

A partial explanation of the previous neglect of small firms stems from earlier preoccupations with the 'big is beautiful' concept that stressed the economies of scale available to large firms in industrial sectors where mass production was prevalent. Certainly, this argument has much truth in industries where only large firms can compete, such as motor vehicles or chemicals (Freeman 1974). However, large-firm economies of scale are not possible in all forms and scales of industrial production, as the level of international small firm employment shown in Table 2.1 clearly confirms. The Japanese example is enlightening in this context. No analysis of the remarkable post-war industrial growth of Japan would be complete without comment on the success of large Japanese corporations. High levels of investment, efficient production methods and a 'cradle to the grave' approach to employees are features that are often propounded as irrefutable evidence of an optimum and ultimate scale of production. However, Table 2.1 clearly shows that 51.6 per cent of the total manufacturing workforce of Japan in 1970 was found in firms of less than 100 workers, the largest national percentage in this size category. These figures would suggest, and observation of the total Japanese industrial system confirms (Anthony 1983), that the success of large Japanese companies could not be achieved without the assistance of small firms.

In particular, subcontracting of both sub-assemblies and finishing work to small firms allows large corporations with strongly binding employment policies to expand, or more importantly, contract, without the trauma associated with hiring or firing. Put simply, the small firm sector absorbs fluctuations in the capacity of the Japanese large firm sector and, ironically, allows the existence of the much-praised employment practices in their large counterparts. In this sense, the Japanese small firm sector is the 'less acceptable side' of the same generally impressive total industrial 'coin' on which the various well-known industrial problems, such as poor working conditions, low pay and lack of job security, are hidden.

However, the supportive value of small firms in *all* Western economies is considerable. Although perhaps to a lesser extent than in Japan, most small businesses interact with their larger counterparts, representing varying levels of either dependency or threat, depending on the manner in which the individual manufacturing activities of small firms complement or compete with the activities of larger firms (Averitt 1968). Here, the distinction between subcontracting and product-based firms, used subsequently in this book, is critical. Subcontract firms perform a largely subservient role and are consequently not a threat to larger-customer corporations. However, product-based firms are frequently in competition with large firms and *will* pose a threat if their product achieves a volume of sales that would either depress the sales of a larger competitor or attract larger firms to a new and profitable market established by the small firm.

Seen in this light, the role of the small firm is either subservient or adversorial, and in the medium term a clash will occur between those small firms which adopt an adversorial stance towards larger competitors. If the imbalance in resources between large and small firms is taken into account, and if the limited power of monopoly-restricting legislation is discounted, it might be anticipated that few of the minority of firms that do grow rapidly will survive the acquisitive attentions of large competitors and become truly large businesses. Moreover, there is evidence from the United States and Britain to indicate that the acquisition activities of large corporations have absorbed many rapidly growing small and medium-sized firms (Smith 1982; Channon 1973; Meeks 1977). Since it is unlikely that the majority of such small firm acquisitions would be undertaken to acquire physical assets, the most probable motive is to obtain innovative new products, for it is clear that in certain areas of industrial production, notably electronics, significant innovation advantages exist in the small firm environment (Von Hippel 1977; Roberts 1977; Townsend *et al.* 1981). However, when cash shortages occur because of the growth caused by successful innovation, large firms are often able to acquire all or part of the equity in such small firms and 'cream off' the accumulated innovation.

A ramification of the absorption of small firms by large corporations has been the proportionate decline in many Western economies of the small firm sector's contribution to industrial output. This pheno- menon has been particularly prevalent in Britain (Prais 1976). Although the periods, nations and definitions of smallness vary, there appears to be a general stagnation or decline of small firms in most Western nations with respect to their share of both manufacturing employment (Table 2.2) and manufacturing output (Table 2.3, Japan excepted). These fragmented data are somewhat confusing if a detailed comparison, where this is possible, is attempted. For example, the United Kingdom appears to have experienced an increase in the share of total manufacturing employment among small firms between 1968 and 1976, albeit by only 2 per cent (Table 2.2). Conversely, the contribution of small firms to manufacturing output in the United Kingdom during this same period appears to have fallen by 7 per cent (Table 2.3). Clearly, these data are not necessarily contradictory and may well be explained by expansions and contractions within individual industrial sectors. For example, labour-intensive sectors with lower levels of output might be expanding while capital-intensive sectors with low labour inputs, but high levels of manufacturing output, are in decline. However, in a period of intensifying depression since the 1973 oil crisis, but especially since 1980, it would be reasonable to expect that the business environment for small firms has not been conducive to growth and prosperity. The tendency for prices to be forced down in the shrinking markets of a recession, and the greater cost and scarcity of investment capital (Bannock 1981), must have made conditions for small *manufacturing* firms less conducive to

Table 2.2 Changes in the share of total manufacturing employment in establishments with less than 200 employees

	Year	(%)	Year	(%)	Year	(%)
Countries with continuous decline						
Germany	1953	40	1963	34	1976	31
Sweden	1950	56	1965	53	1975	41
Norway	1953	70	1963	65	1975	58
France *	1954	58	1963	51	1975	41
Countries with increases and decreases						
UK	1954	33	1968	29	1976	31
Switzerland	1955	66	1965	61	1975	64
Canada	1955	46	1964	47	1975	44
USA	1954	37	1963	39	1972	38

* Small establishments in France, although classified as having experienced continuous decline, increased their share of employment between 1963 and 1965.

Source: Bannock (1981); Bolton (1971); Storey (1982).

Table 2.3 Small and Medium-sized Enterprises: their changing share in manufacturing output in four countries

	Share in manufacturing output		
France		1970	1976
(<500 employees)		29.6%	28.3%
Japan		1962	1974
(<300 employees)		48.4%	51.3%
Ireland		1963	1968
(<£200,000 turnover)		21.2%	13.6%
UK	1951	1968	1976
(<200 employees)	32%	25%	approx. 18%

Source: Rothwell and Zegveld (1982).

their establishment and growth. While Ganguly (1982) has recently indicated a modest increase in the establishment of new British manufacturing firms in 1981, several qualifications on the new method by which the data were compiled must cast doubt on whether this information is a true reflection of an improvement in small firm prosperity.

A summary of this, admittedly fragmented and partly contradictory, evidence must conclude that, generally, international small firm growth in manufacturing has stagnated in terms of both its absolute and its proportionate contribution to national employment and output. This overall picture perhaps reflects the general decline of manufacturing industry and the recent recession in many of the countries concerned. Indeed, the exception of Japan to this general pattern might be explained by the relative success of the Japanese economy in recent years.

2.2 Sub-national regional variations in the contribution of small firms to manufacturing employment: the British case

The preceding evidence has indicated that small firm employment is a significant part of any national industrial employment. The approximate range of contributions from the 20 per cent to 50 per cent level in different national economies might be respectively interpreted as important and critical. But intuition suggests, and the following evidence confirms, that any national level of small firm employment will vary when it is broken down into its regional components. In this sense, the national figure can be seen as aggregated regional means.

The propensity of a national economy to generate new firms must be, at least in part, dependent on the overall economic buoyancy of the nation concerned. However, this argument may also be applied at the sub-national regional level. There is no doubt that British regional industrial performance has been unbalanced for many years and that a distinct core–periphery pattern exists between the South East and the development regions of the North and West (Keeble 1976; Cameron 1979). If the negative influences of older industries in the development regions are taken into account, both in terms of their contraction in recent years and in terms of the tendency for these industries to be dominated by large scales of production and large plant size not conducive to small firm 'spin-off' (Johnson and Cathcart 1979), it might be expected that the development regions would possess a lower proportion of small firms than either the national average in general or the South East in particular. Due to problems concerning the way in which the Department of Industry presents data through HMSO, it is impossible to obtain direct, regionally disaggregated data on small firms, or 'enterprises' as they are termed in the statistics. None the less, Table 2.4 provides a regional breakdown of employment in establishments of less than 50 employees. This unusually small size band has been chosen in order to maximise the concomitance of small establishment size with independent status, since membership of multi-site corporations is less likely in this category. While the proportion of employment in most regions in plants of less than 50 workers falls close to the British average of 15.7 per cent, there are two obvious exceptions. The prosperous South East achieves a 20.6 per cent level of employment, while the North records the lowest proportion of 9.3 per cent. Clearly, in the South East, employment in establishments of less than 50 workers is fractionally greater than twice the figure for the North. Wales registered a poor 12.6 per cent level, while in Scotland, the remaining development region, the level

Table 2.4 Employment in small establishments in Britain by region

Region	Plants employing <50 workers as a % of total regional manufacturing employment	Employment in plants employing <50 workers	Total employment
North	9.3	39,095	418,973
Yorkshire & Humberside	14.9	102,908	688,001
East Midlands	15.7	88,151	562,908
East Anglia	15.4	29,956	194,703
South East	20.6	353,197	1,711,443
South West	15.3	63,907	416,579
West Midlands	13.4	126,487	943,423
North West	13.9	135,352	976,960
Wales	12.6	38,614	306,789
Scotland	15.3	87,744	572,466
Britain	15.7	1,065,411	6,792,245

Source: Department of Industry (1978).

is almost equal to the national average. This relatively good performance by Scotland may be explained by political and structural factors, and these are examined in greater depth in subsequent chapters. However, the dominance of the South East is clear.

Ostensibly, these data are open to the 'so what' criticism in which it might be argued that the size of plant or firm in which workers are employed matters little, the critical factor being the job itself. However, it should always be remembered that any single set of statistics for a given year are merely a time slice through an evolving process of employment generation and decline shared between different forms and sizes of industrial production. Although no comprehensive regionally comparative data exists, data bases that have monitored the growth performance of small firms in the medium to long term have observed their contribution to be significant.

Fothergill and Gudgin (1982), working on a data base for Leicestershire in the British East Midlands standard planning region, discovered that small firms founded since 1945 accounted for 17.0 per cent of the county's employment in 1975, with an estimate of 23.0 per cent projected for 1979 (Table 2.5). These data suggest that a healthy small firm sector is a major contributant to the long.-term employment growth of any region. Hence, the regional pattern indicated in Table 2.4 does not auger well for the North's industrial employment growth as generated from the small firm sector. Such a deficiency puts added emphasis on the employment capacity of the region's large-plant-dominated sectors which are often shedding labour in an attempt to become competitive with foreign firms, as in the cases of the shipbuilding and iron and steel industries. None the less, given an unbalanced pattern of regional small firm employment, and the implications for future employment growth that these distortions bring, such trends are only the surface symptoms of a deeper cause. Regional differences in the establishment and subsequent employment capacities of firms stem directly from the success achieved in creating and selling goods or services in the market (Feller 1975). Innovation is the key to the successful establishment and fast-growing employment capacity of small manufacturing firms (Freeman 1974; Rothwell and Zegveld 1982). Hence if innovation *is* critical to subsequent small firm prosperity and employment growth, it might be expected to show a similar pattern to the regional distribution of small firm employment.

The preceding contextual outline is now enlarged by conceptual consideration of the regional implications of technological change and

Table 2.5 Employment in new firms in Leicestershire, 1947–1979

	Cumulative employment in post-1947 new firms	As % manufacturing employment
1947	0	0
1956	6,100	3.8
1968	14,800	8.8
1975	27,600	17.0
1979 *	36,000	23.0

* Estimate.

Source: Fothergill and Gudgin (1982).

innovation in the small firm, followed by hard evidence on the manner in which small firm innovation levels do conform to a regional core–periphery model in a similar manner to regional small firm employment.

2.3 Small firm innovation and regional growth: some basic themes

Technological change, which is the end-product of the innovation process, has for many years been known to be a major spur to economic growth (Solow 1957). The rebirth of interest in Kondratiev cycles (Kondratiev 1935) stems from a recognition that, in contemporary industrial conditions, the concept of cyclical bursts of innovative activity is as useful in explaining the impact of the microprocessor as it was in getting to grips with the wide-ranging effects of the steam engine on Victorian economies (Rothwell and Zegveld 1982). There has long been an acknowledgement of the influence of technological change on economies at an abstracted theoretical level (Denison 1967; Schmookler 1972), while other work has asserted the importance of technological change in explaining the relative industrial performance of nations (Boretsky 1975), or of firms (Sappho 1972). However, it is only in recent times that spatial variations in technological change have been proposed as causes of differential regional growth at a sub-national level (Thomas 1975; Thwaites 1978). While proof of this assertion would clearly involve an attempt to measure hypothesised regional variations in technological change (a challenge met in the following sections), it is sufficient in the current context to note that regional variations in industrial growth in all sizes of industrial firms owe much to spatial variations in the firms' innovation performances.

2.4 Regional growth through small firm innovation

Intuition, and evidence on the progress of great enterpreneurs who transformed regions through small firm growth based on product innovation, clearly show the potential for the long-term impact of small firm innovation on regional employment levels. Henry Ford and William Morris were giants of the motor-vehicle industry who, mainly through their own innovation and resolve, created industrial sub-regions where previously little industrial employment had existed. Today, this entrepreneurial phenomenon remains important, although the motor-vehicle industry, the growth industry of the early twentieth century, has been replaced by other industries such as electronics, a key growth sector in the current world economy. In electronics, the names of Hewlett and Packard in the United States and Sinclair in Britain are the modern equivalents of Henry Ford, since they restate the potential of entrepreneurial-led small firm growth founded on innovation rooted in the leading-edge technology of the day (Morse 1976).

However, the innovative contribution of the great entrepreneur to regional small firm employment growth is merely a glamorous veneer on the substantial contribution, noted by Fothergill and Gudgin (see Table 2.5), of many small firms with more modest rates of innovation and subsequent employment growth. Moreover, since the principle of innovation in small firms remains the same for successful and unsuccessful small firms alike, it is useful here to consider this principle and its hypothetical regional impact in advance of the data in this chapter on variations in regional product innovation levels, and as a general precursor to the subsequent empirically based chapters on detailed aspects of the innovation process.

There are two fundamental means by which technological change can be achieved in small firms, and these two processes may have markedly different regional growth potentials. Technological change is achieved either by process innovation or by product innovation. These two forms of technical progress are considered in turn, together with their differing regional impacts.

2.4.1 Process innovation

At best, the impact of process innovation on small firm growth, and certainly on employment, can be termed marginal. Basically, the relevance of process innovation to small firms is reduced because savings in process costs are most frequently linked to mass production, which is not a scale of output at which, by definition, small firms operate. The introduction of new process techniques at the small firm level predominantly takes the form of new machinery. Since most small firm production methods range from one-off assemblies to small batch, there is little scope for expensive high-volume process machines to pay their way through a reduction in unit costs. It is true that the latest computer numerically controlled machines (CNC) are highly flexible, both in terms of function and in volume of output. But while the technical advantages of such possible acquisitions are clear in quality and capability terms, the economics of operation are frequently such that the machine cannot 'pay for itself' due to the fragmented nature and low volume of small firm production.

Perhaps the greatest scope for the introduction of new process machinery lies in the small-firm-dominated field of subcontracting, where the necessity of low unit costs is a feature of the strong competition experienced in this field of industrial activity. However, the majority of small subcontracting firms are forced to operate in areas of production where low volume renders the manufacture or finishing of certain components uneconomic for a larger client firm. If the volume of production for any subcontracted item becomes large over a period of time, the client firm may well decide to manufacture the item internally, and vertically reintegrate production. Unfortunately for the subcontracting firm, this reintegration often happens at the very moment when expansion due to increased volume begins to make high-volume process machinery economically viable. The subcontract printed circuit board industry examined in later survey material displays many of these features.

If the firm has acquired several customers, the prospect of introducing mass production machinery is often inhibited by the diverse technical nature of such a 'mixed bag' of jobs, which requires reorganisation of production methods and resetting of machines, and again reduces the advantage of relatively high-volume process machines. Generally, process innovation is not an area of technology where small firms gain major competitive advantages that will enable expansion and employment growth. Indeed, the clear implication is that employment levels are likely to be decreased by the utilisation of new process machinery, and the inapplicability of such techniques to small firm production methods might enhance, or at least protect, the sector's medium-term share of national employment.

Indeed, on a regional level, the larger proportion of employment in small firms in the South East (Table 2.4) might offer greater protection from unemployment due to the reduced applicability of process machinery to this scale of production. Conversely, the large-plant-dominated mass production industries of the development regions are, on balance, more likely to introduce new process technologies with subsequent potential for job losses. Certainly in iron and steel, coal mining, and textile production, job losses through process innovation have been consistently evident since the end of the Second World War. Moreover, recent research has indicated that process innovation levels in the development regions are similar to those in the South East of England, unlike product innovation levels, where the development regions are laggards (Oakey *et al.* 1982). Thus, there may well be a neutral or inverse relationship between process innovation and regional employment growth, mainly enacted through the large-plant sectors of the economy and with particular scope for applicability and subsequent job losses in the development regions. However, as already stated, this is a process mainly outside the sphere of small firm operation.

2.4.2 Product innovation

Although, as will be noted in Chapter 3, not all small firms rely on product manufacture since many offer services based on sub-contracting of various kinds, successful product innovation is a major spur to rapid employment growth in small firms. Clearly, such employment growth largely depends on the rate at which sales of a given product increase, and this increase depends on two major factors. First, the price of the product will strongly affect its level of demand in a fundamental economic sense. But, second, price will be less important if the product is at the leading edge of technology and boasts a sophisticated or unique technical specification. Indeed, improvements in specifications are the main means by which small firms gain advantage in competitive conditions (Oakey 1981). An amalgamation of both low price and high technical specification would obviously be a recipe for success, although such a combination is rare, and perhaps more important, not essential for reasonable profit and subsequent growth.

A large number of product-based firms are currently emerging in the high technology electronics family of industries. These industries are particularly conducive to the formation of small firms since the advancing new technologies are constantly creating new opportunities for specialist sub-component manufacture. Because these product niches are new and specialist in nature, they do not sustain the level of production that would be attractive to large firms, at least in the early stages of production. The economic viability of production at this low volume of manufacture is ensured by the high technology expertise embodied in the output of these small firms by internal research and development workers. The efforts of these researchers ensure that the sophisticated specification of the product will command a high price in the market. However, if large production runs become feasible at some later date, it is likely that larger firms will move into these small firm production niches and take over either the firm itself, or much of the small firm's emerging market (Averitt 1968; Channon 1973). Hence, small firms are forced into, and sometimes deliberately choose, specialist niches in rapidly evolving technologies. None the less, even in specialist markets, profits can be substantial due to the high prices charged which, in turn, ensure rapid growth for certain firms. Moreover, it is possible that small firms can occasionally break through the restrictions of specialist market niches and the competitive pressures of large firms to become very large firms in their own right, as evidenced by Fairchild and Texas Instruments in the United States (Morse 1976).

At the regional level, a high incidence of product innovation in small firms ensures the survival of the business at the very least, and at best heralds future employment expansion and growth. There are few areas of production where technical progress is totally static. Hence there is a variable need for innovation directed at new and improved products which will bring market advantage to individual firms within a region. While at the extreme, high technology small firms may need to improve radically the specification of products every five years, most firms will periodically need to improve existing products if market share is to be maintained. Unlike process innovation, a product innovation is often the *raison d'être* of a small firm, and the confidential 'in-house'knowledge implicit in the conception and construction of a product may be the only means by which the price and market niche of a firm's product can be maintained in competitive national or international conditions.

Unlike subcontracting firms, businesses based on evolving products are not usually dependent on a single customer and are more likely to export their products outside the region or abroad. This ensures that, although small product-based independent firms are likely to source inputs locally, with important multiplier benefits to the local economy, they are less likely to be dependent on a large single local customer, as compared with their subcontracting counterparts (Storey 1982). Moreover, it is clear that their scope for evolution into large firms is enhanced by the possession of a highly competitive product. Evidence from the semi-conductor industry of the San Francisco Bay area of California indicates the employment growth that can be achieved from a series of product-based small firms (Rothwell and Zegveld 1982). It is evident that the evolution of such product-based small firms into large corporations ensures that they in turn act as significant 'incubators' for further product-based small firms, with obvious cumulative advantage for the region concerned.

Put simply, when product innovation is compared with process improvements in the small firm context, product innovation is a much more potent measure of the vitality and internal technical sophistication of small firms. Hence, given the preceding comments on the stimulative effects of product innovation on small firm growth potential, it might be argued that the higher levels of small firm employment in the prosperous South East of England (Table 2.4) reflect a higher level of small firm product innovation in this region, when compared with the rest of Britain. Until recently, such speculation has remained unsubstantiated. However, new data on the regional innovation performance of small firms indicate that significant differences in innovation performance *do exist* between the prosperous regions and the development regions, and such differences must exert influence on the regional employment variations observed in Section 2.2.

2.5 Regional variations in small firm product innovation levels

The following data have been selected from work performed since 1978 at the Centre for Urban and Regional Development Studies at Newcastle University. The specific results presented here for small firms are lent added weight by the broadly based objectives of the original survey. The remit of the research was to investigate and measure the extent of regional variations in the innovativeness of British manufacturing industry in general. The stimulus for this work had emanated from the final report of the Northern Region Strategy Team, in which it was argued that the industrial structure of the Northern region and its general industrial environment were not conducive to innovation (NRST 1976). These general impressions were set in a more formal academic context by Thwaites (1978), which acted as a basis for the hypotheses tested in the research summarised below. However, it is important to stress that no particular concern was directed towards small firms at the outset of the research. The consistent pattern of a poorer innovation performance by small firms in the development regions when compared with those in the prosperous South-Eastern 'core' of Britain emerged from an analysis of all sizes of manufacturing firm.

Initial research designed to expose regional variations in manufacturing innovativeness at plant level was based on a combination of published material, which identified firms receiving the Queen's Award to Industry for innovation, and data kindly provided by Joe Townsend at the Science Policy Research Unit (SPRU) at Sussex University. The SPRU data base contained evidence, carefully derived from a panel of industrial experts, on significant British innovations (Oakey 1979a). The two data sets were merged and duplicate innovations removed. The final data set comprised 383 innovative firms detected over the period 1965–78. These firms were sent a questionnaire designed to pinpoint specific factors such as the development location of the innovation (if not at the recorded address), the first site at which the innovation was introduced into production (if not at the recorded address), and the nature of the innovative firm's corporate organisation. Of the 383 firms contacted, 323 (86 per cent) responded, giving a reasonably comprehensive set of information on the nature of *most* of the significant innovations introduced in a broad spread of industrial sectors during the study period.

While other valuable results on multi-site corporate development and production location strategy for these predominantly product-oriented innovations were discovered, the examination of small establishments is the central concern of this chapter. The analysis of Table 2.6 again adopts the generally safe procedure of equating small plant size with independent status. The results indicate that, based on an innovation index designed to control for regional plant size structures, in the West Midlands and South-Eastern planning regions (the most prosperous parts of Britain during the survey period) small plants were over twice as innovative as their development region counterparts (Oakey 1979a).

Conversely, the difference between the prosperous regions and the development regions, after control for total numbers of plants in the larger size bands, was insignificant. An interpretation of these results argued that since larger plants in the development regions were predominantly branch plants of a wider multi-site corporate organisation, they were less dependent on their local area for innovation. Clearly, the transfer of a new product innovation for production in a development region branch plant from a headquarters in a prosperous area would directly aid the development region's apparent 'product innovation' performance as shown in Table 2.6. However, in many ways this would be a false picture, since the critical research and development involved was not *indigenous* to the recipient development region.

Table 2.6 Innovation performance compared with size structures for selected British planning regions

Size of plant (workers)	Development regions			South East and West Midlands			
	A	B	C	A	B	C	D
1–99	10	12,287	1,229	37	19,994	540	2.3
100–999	38	3,939	103	67	6,541	98	1.1
1,000 and over	34	348	10	39	383	10	1.0

A = Number of innovative plants.
B = Number of plants in region.
C = Number of regional plants required to produce one innovation.
D = Coefficient of regional innovative difference, i.e. (Development regions, col. C) divided by (South East and West Midlands, col. C); unity = no difference.
Source: Oakey (1979a).

In instances where the total level of development region innovation is low, the influx of a small number of innovations from outside the region (from production or research and development locations in other parts of the nation or abroad) to corporate branches may make a significant difference. Moreover, apart from the obvious instance of product innovation transfer, the wider resource environment of plants that are part of a national or international network of production will provide subtle supportive advantages to in-house innovation in the plant concerned. This point is central to the major theme of this book, since it is the ability of branch plants in development regions to bypass the often locally poor innovation environment and utilise a system of corporate resources, such as organisational expertise, centralised research and development, and corporate investment capital, that provides branch plants in a development region with a *major* advantage over the small, independent single-plant firms considered in this book. In this sense, branch plants can be viewed as taking advantage of all the positive benefits available to production in the development regions, such as government grants and cheaper (often female) labour, *without* suffering the negative effects of a region with little research and development infrastructure, a declining banking and insurance sector, and a generally poor environment for indigenous industrial innovation and growth. Thus, it can be argued that the poorer performance of small independent firms located in the development regions is directly linked to a poor local environment for innovation.

This argument is further strengthened by more recent detailed data on both the regional innovation performance of predominantly small, independent single-plant firms and the external assistance obtained towards product innovation by development region plants that are members of multi-site corporations. The following results emanate from research that sought broadly based measurements of regional variations in British manufacturing innovation (Oakey *et al.* 1982). The fieldwork focused on the major determinants of innovativeness: research and development, process innovation, and the introduction of new products. Analysis was based on an extensive postal questionnaire which drew information from 807 plants nationally, and on detailed interviews in 174 establishments in the British standard planning regions of the North, North West and South East. (For further details see Thwaites *et al.* (1981).)

The strongest regional differences in innovation levels between the core South-Eastern region and the development region periphery related to product innovations introduced during the study period 1973–7. Moreover, the most clearly defined result of the whole research, contained within this overall pattern of product innovation, was the regional difference in product innovation in predominantly small, independent single-plant firms. Table 2.7 clearly indicates that the level of product innovation in this category of Northern region survey firms was 55 per cent compared with 85 per cent in the comparable South-Eastern group. It is relevant to earlier arguments that the development region plants that were members of a corporate group recorded an 87 per cent level of product innovation, only four percentage points lower than similar plants in the South East of

Table 2.7 Incidence of product innovation by plant status, 1973–1977

Regional groupings	Single plant						Group plant					
	Innovation		None		Total		Innovation		None		Total	
(N = 786)	N	(%)	N	(%)	N	(%)	N	(%)	N	(%)	N	(%)
Development areas	23	(55)	19	(45)	42	(100)	59	(87)	9	(13)	68	(100)
Intermediate areas	99	(75)	33	(25)	132	(100)	146	(87)	21	(13)	167	(100)
South East	157	(85)	27	(15)	184	(100)	176	(91)	17	(9)	193	(100)
Total	279	(78)	79	(22)	358	(100)	381	(89)	47	(11)	428	(100)

Source: Oakey *et al.* (1982).

England. Given that there was little evidence of similar regional variations in process innovation in single-plant or 'group' firms (Oakey *et al.* 1982), explanations for the sharp regional variations in product innovation in single-plant firms were based on the relative confidentiality of product innovation, as compared to process innovation, in industrial firms. Most product innovation was in-house, relying on local resources, while most process innovation took the form of machinery purchased from outside the region. Again, these results imply that product innovation is the best measure of both plant-level and local regional innovation performance.

Other research has shown that, in the short term, product innovation is a highly confidential activity since it is the means by which firms gain advantage in the marketplace (Oakey 1981; Rothwell and Zegveld 1982). Hence most product innovations tend to be developed in-house, and this is confirmed by the figures in Table 2.8. Given the earlier findings (in the previous analysis of significant

Table 2.8 Incidence of product innovation developed in-house

Region	Innovation		None		Total	
(N = 136)	N	(%)	N	(%)	N	(%)
North	17	(53)	15	(47)	32	(100)
North West	37	(77)	11	(23)	48	(100)
South East	41	(73)	15	(27)	56	(100)
Total	95	(70)	41	(30)	136	(100)

Source: Oakey *et al.* (1982).

British innovations) on the advantages of corporate links with group plants outside the development regions and their positive impact on development region innovation, intercorporate transfers might be perceived to play an important part in the aggregate development region innovation performance. It is certainly true that Table 2.8 clearly shows a much inferior level of in-house product innovation in the North. The 53 per cent level of in-house product innovation in the North is much lower than the 77 per cent and 73 per cent levels achieved in the North West and South East respectively. Moreover, Table 2.9 confirms that the lower level of in-house product innovation in Northern region plants is mainly caused by the influx of innovations to group plants from external sources, later discovered to be from outside the region or from abroad (Oakey *et al.* 1982). For example, twelve of the fifteen externally derived innovations in the Northern region sample were attributed to group plants.

Table 2.9 Incidence of in-house product innovation by plant status

Region	Single plant			Group plant		
	Innovation	*None*	*Total*	*Innovation*	*None*	*Total*
(N = 135)	N *(%)*	N *(%)*	N *(%)*	N *(%)*	N *(%)*	N *(%)*
North	7 (70)	3 (30)	10 (100)	10 (45)	12 (55)	22 (100)
North West	12 (80)	3 (20)	15 (100)	25 (76)	8 (24)	33 (100)
South East	16 (76)	5 (24)	21 (100)	24 (71)	10 (29)	34 (100)
Total	35 (76)	11 (24)	46 (100)	59 (66)	30 (34)	89 (100)

Source: Oakey *et al.* (1982).

Hence the performance of group plants in the development regions was boosted by the acquisition of product innovations from sources external to the region. Clearly, the absence of this opportunity for small independent firms in the development regions, together with their previously mentioned greater dependence on a locally poor innovation environment, must in part explain their lower level of innovation. However, it should be re-emphasised that the enhanced performance of development region group plants is misleading if taken at face value, since the acquired innovations do not necessarily stem from the realisation of *indigenous* regional innovation potential.

2.6 The regional industrial environment, innovation and employment: a summary

This chapter has discussed the importance of the small firm in national and regional contexts with particular attention to the employment contribution of small firms to these differing spatial scales. It has been noted that the more prosperous South East of England possesses a higher proportion of small firm employment than the development areas. Subsequent hypotheses and supporting data have indicated that the higher level of small firm employment in the South East reflects a higher level of product innovation, which must be linked to prosperity and employment growth in small firms. This argument leads to the main hypothesis of this book, which argues that the evidence of a consistently poor level of product innovation in small independent firms in peripheral development regions is largely due to the poorer local resource environment.

It has been important to establish at this early stage the background and broad objectives of the research on which this book is based in order to give a clear sense of the overall purpose of the work. However, while Chapters 5 to 9 present a detailed examination of specific components of the resource environment of small high technology firms through an amalgam of relevant theoretical and empirical evidence, Chapters 3 and 4 continue the contextual themes of this chapter by considering, respectively, the impact of industrial structure on the establishment and growth of small firms, and the general theoretical principle of agglomeration.

Clearly, there is a strong link between these two themes in the sense that the industrial structure of a region, defined as its mix of production activities, largely determines the potential for agglomeration economies in small firms. However, the main consideration of industrial structure in Chapter 3 is to offer a broad conceptual picture of how the existing industrial structure influences the type and number of small firms generated in a region. Chapter 4, however, introduces the pervasive theoretical concepts of agglomeration and product life cycle as a precursor to more detailed utilisation of these general principles in the subsequent and specific empirically based chapters. The theory of agglomeration merits a treatment in these introductory chapters because its rationale is a useful theoretical articulation of the arguments in this chapter on the varying quality of local resource environments. In particular, it is a useful tool in seeking to explain the emerging concentrations of high technology industries in, for example, the Silicon Valley area.

3 The effects of industrial structure on regional small firm growth

3.1 Industrial structure and regional growth

The total industrial output of any industrial area, whether defined on national, regional or sub-regional scales, will be comprised of a mix of many forms of expanding or contracting industrial production. This mix varies in the medium to long term to alter the industrial structure of the region and its relative prosperity. In the short term, however, industrial structure is the source of much inertia and, particularly in regions with declining industries, it speaks more for the past success of a region than of future economic potential. History shows clearly that the locations of the 'boom' industries of one era may well be the foci of tomorrow's problem regions. Most of Britain's industrial areas were developed on the basis of a propulsive industry such as coal, iron and steel or textiles. At their peaks, the opportunity costs of producing such outputs in these now frequently depressed regions were so overwhelming that they precluded other potential forms of industrial activity, thus creating a series of specialised industrial agglomerations in Scotland, the North East and Wales.

In the case of the North East of England, steel and shipbuilding might be cited as rationales for such industrial concentrations. These industries grew rapidly on the basis of good profitability and quickly reached sizes that dominated the total industrial structure of the region in terms of both the sectors concerned and their component firms. The products of these industries matured, and technological emphasis moved from perfecting the product to increasing the efficiency of production through the introduction of new process techniques. However, at some point the drive to reinvest and to refine production techniques began to decline (NRST 1976). Whether this decline in investment was caused by reduced sales resulting from spontaneously developed foreign competition, or whether foreign competition was attracted to compete in the markets of these overmanned and technologically stagnating industries is debatable. But once the decline in reinvestment began to occur a downward spiral of investment deficiency was initiated, and subsequent vicious circles of investment failure amplified the results of past neglect (Figure 3.1). When industries have been pitched into this downward spiral of decline, there is little hope that the 'free market' will offer any means of redemption. Falling profits inhibit reinvestment, and a failure to reinvest causes high production costs relative to those of other producers who *are* reinvesting, thus resulting in a further reduction in profits due to reduced sales of overpriced and uncompetitive output.

However, it would be foolhardy to discount complacently the problem of the collapse of regions with monolithic industrial structures as a lesson of economic history. The West Midlands of England currently shows many of the symptoms of a specialised industrial problem region in that the major industrial activity of motor-vehicle manufacture is in decline in both output and employment terms. The relatively recent prosperity of both the sector and the host region confirms that there is never any guarantee that a currently prosperous industry will remain profitable in the future. Any economy that is dominated by a single industry is always fragile since it can be devastated by a decline in demand for the product concerned, or by cheaper competition from outside the region. Hence, it is desirable that such regions attempt to achieve a broader spread of manufacturing activities.

While the diversification of regional industrial structures is the key to reducing the trauma of a sudden decline in any given sector, such an objective is, in practice, difficult to achieve. In seeking regional sectoral diversification, the unattractive nature of the job market, in areas where declining industries predominate is a more intransigent problem than the physical infrastructural inertia caused by the unadaptability of purpose-built production facilities such as shipyards or steelworks. While new factories can be erected in a matter of months, changes in lifetime skills and in social attitudes towards work may take decades to accomplish. One of the many vicious circles that appear to exist in regions with industrial structures dominated by declining industries is the phenomenon of an unattractive mix of labour skills (such as those of steelworkers and shipyard welders) repelling modern expanding manufacturers from home or abroad who seek, for example, skilled blue-collar electronic technicians or white-collar research and development personnel. Because the potentially migrant firms are deterred by the paucity of local skills, there is little chance of accumulating the sought-after skills in the region concerned (Oakey 1983).

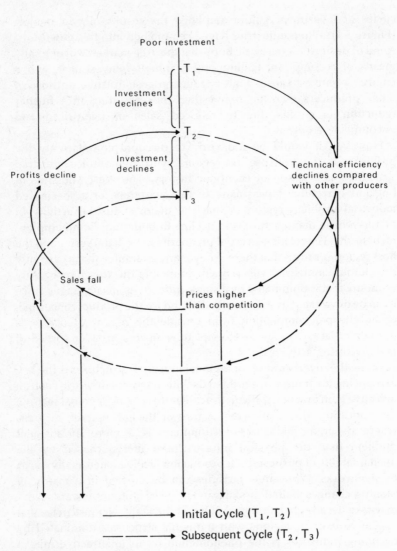

Fig. 3.1 Cycles of regional investment decline

There can be little doubt that a poorly diversified local economy, once established, significantly inhibits subsequent industrial structural change, due to the repulsion of possible new forms of production from outside the region, and the limited potential for indigenous industrial change from the existing mix of production. However, if these problems are retraced to their origins, any explanation of the *initiation* of the process of sectoral decline must address the detailed investment deficiencies in the behaviour of management towards technological change in the industries concerned. There are two main, technologically based, reasons for the decline of a large dominant regional industry. First, the products of an industry may become wholly or partly obsolete. Seen in its most acute form, this phenomenon is typified by the experience of steam railway engine or mechanical watch manufacturers. Both these industries experienced an exhaustion of demand for their products that was virtually instant in nature. However, individual firms that are driven to bankruptcy in this manner suffer from the ramifications of a previous neglect of investment in research and development directed at new product development. The demise of the British motor-cycle industry is a good example of this phenomenon. A second major reason for industrial decline is the failure, in industries with relatively mature or standardised products, to reinvest in new process machinery and construction technologies. Process investment in these industries is important due to the need for production cost reduction to create price advantages in very competitive markets (for example iron and steel, and textiles). Unlike the case of product obsolescence where the effects are relatively sudden, the impact of decline stemming from outdated processes is generally medium to long term. Indeed, it is often impossible to examine declining regional industry and isolate the precise year in which the decline began; but looked at over decades, the decline is readily apparent.

It may not be initially clear to the captains of the declining industries concerned that problems exist. A disbelief that new entrants to the market can survive, let alone prosper, is often prevalent. An inability to complete on price is frequently explained away by the supposedly inferior quality of a competitor's products. But slowly, apparently precocious entrants to the market take progressively larger shares until they cannot be dismissed and must be taken seriously. While the industries of Japan have often been cited in this context, other more contemporary examples in South-East Asia can now be identified, such as the growing industrial producers of South Korea and Taiwan.

On other continents, India and Brazil are further examples of countries with corporations that enjoy labour costs that compare favourably with those of Japan. Indeed, many of these producers now also use the latest process machinery to complement labour cost advantages. And if Japan itself is under pressure in industries it has progressively dominated since the end of the Second World War, the prospects are not good for Western regional economies with a high proportion of employment in many of the sectors concerned such as textiles, shipbuilding and iron and steel.

In a British development region context, it is obvious that many of these declining industries are those that have dominated the indigenous output of such areas for at least a century. High wage costs caused by a failure to reinvest in labour-saving capital equipment have precipitated the loss of markets and have caused the severe contraction of these industries. Without extensive government assistance and protective tariffs, whole industrial sectors in the depressed regions of Britain would have been wiped out. The employment losses resulting from radical declines in regional production would have focused unbearable economic and social stresses on these regions' economies. Such regions require radical measures to break the circle of decline simplified in Figure 3.1. However, since the decline of once-prosperous industries in depressed regions can be largely attributed to a failure to reinvest in the development or acquisition of new technology, the reversal of decline may also be achieved by a solution based on technological change through vigorous investment in product and process improvement (Thwaites 1978).

There are two main dimensions to a technology-based regional policy for the regeneration of depressed industrial regions dominated by large declining industrial sectors. First, investment must be provided to make the existing, predominantly large, declining industrial sectors more competitive through the replacement of obsolete processes and outdated working practices, thus enabling increased output from fewer operatives, resulting in reduced prices and an improvement in competitiveness. Indeed, since the Second World War great strides *have* been made in the British coal mining, textiles and steel industries. There is no doubt that such measures have at least slowed the rate of output and employment decline in these sectors. But these mainly process-oriented technological improvement are of reduced importance to the present discussion, not only because their value to industrial growth has been known for many years, but also because such technological improvements are mainly applicable to very large firms operating in mature industrial technologies and are thus of peripheral interest to this book on small high technology firms.

However, the second potential dimension to a technology-based regional policy *is* central to several themes in this book. In order that regional industrial structures should be diversified to avoid problems of the past associated with dominant industries liable to subsequent decline, it is essential that new, diversified forms of industrial employment should be created in depressed regions, both through inward investment in new production, and as a result of the stimulation of the generation and growth of new small firms. Most of the attempts to achieve development region industrial diversification in the post-war period have centred on mobile industry policies (Keeble 1976). These policies were moderately successful in terms of employment generation in the short to medium term, but their long-term ability to provide secure employment has been seriously questioned (Sant 1975; Smith 1979). In any case, mobile industry policies have been largely negated by the more recent dearth of expanding industry in the wake of the 1973 oil crisis and subsequent recession.

However, in many Western countries much recent stress has already been placed by policy makers on the indigenous potential of regions and their ability to generate new industrial growth through the exploitation of local internal resources (Thwaites 1978; Rothwell and Zegveld 1982). It has been argued that governments should act to mobilise the indigenous potential of regions to encourage growth from within (Ewers and Wettmann 1980). Clearly, the encouragement of new, indigenous, small firm growth must be an important component of any attempt to realise indigenous regional potential. However, the preceding discussion suggests that variations in regional industrial structure will have marked effects on the propensity for small firm generation and growth.

The following sections will focus on specific aspects of the manner in which the generation and innovative potential of small firms are influenced by regional industrial structure. This approach develops from a fundamental consideration of the birth of small firms, which becomes the basis for a discussion of the strong influence of industrial structure on small firms' innovation and growth potentials.

3.2 Small firm birth, growth and innovation

At a conceptual level, new and existing firms are entities that exist at different points on this same small firm continuum, fixed in different positions by their relative age. In this sense an 'established' firm is merely a new firm that has existed for a given number of years and passed through some age cohort category. The subsequent chapters of this book will evoke the concept of the product life cycle to emphasise that many of the problems of small firm birth, such as product innovation and the acquisition of investment capital, are not unique events but reoccur during the life of the firm. Indeed, problems of definition surrounding the difference between 'new' and 'established' firms perhaps reflect the arbitrary nature of such a distinction (Mason 1983). In practical terms, the critical difference between new and established firms is in accumulated experience and resources. However, such experience may be a 'two-edged sword'. On the one hand, there is no doubt that most entrepreneurs develop many diverse skills as their business grows, but on the other, it is also true that such accumulated experience may have an ossifying effect on the dynamism of small businessmen (Boswell 1973; Senker 1979). Clearly, a major advantage of the new small firm is the ability of executives to start trading with a 'clean sheet', free from bad business habits. Moreover, the attraction of organising the business to suit the personal tastes of the founding entrepreneur is a strong motive for the 'spin-off' of executives from larger, less flexible organisations (Shapero 1980). It is also likely that such flexibility is particularly conducive to successful innovation (Von Hippel 1977; Roberts 1977). However, leaving the ambiguous benefits of experience aside for the moment, success in the choice of activity involves a combination of two basic skills on the part of the decision maker (or makers). These are:

(i) business acumen, and
(ii) technical ability.

In most practical instances, a new or existing small firm entrepreneur will possess a mix of both abilities, although business acumen will clearly be minimal in many new small firm entrepreneurs. Business acumen may be all important in certain areas of activity, especially in areas of manufacturing industry where the technology is well established and profit margins are low (such as garment making and printing). In such instances business acumen, reflected in optimal purchasing, contract quoting and the organistion of labour, may be essential to ensure success. In this context technical ability is less important since the technology of the product is established. However, the lack of technological barriers to entry implicit in the generally available technical specification of the product means that the resultingly large number of competing producers force prices down. However, the converse is true of high technology forms of production. In these industries technical skill may far outweigh business acumen since technical barriers to entry preclude other producers who may have higher business acumen but poor technical ability. Hence high prices may be charged for goods and services that stem exclusively from the personal technical ability of the owner (or owners) of the firm. Thus, inefficiencies in business acumen may be masked by the high prices such activities can command. This principle lies at the heart of the economic viability of much high technology industry in general and many high technology small firms in particular.

These arguments are similar to those put forward by Allan Pred (1967) in which he advanced ideas on the location decisions of entrepreneurs. With apologies to Pred, his behavioural matrix can be easily adapted to the present consideration of the dynamic attributes of small firm entrepreneurs. Figure 3.2 represents a matrix into which all small firm entrepreneurs might conceptually be placed. As argued previously, a mix of these two abilities exists in all entrepreneurs. Figure 3.2 shows three main types of businessman. First there is the small firm owner with poor business acumen and little technical ability (represented by the unshaded circles in Figure 3.2). He has little business acumen to help him reduce costs and his inability to obtain protection and higher prices through the development of high technology products means that he must compete in an area of production with tight profit margins. Although it is not inevitable, firms associated with this type of entrepreneur are most likely to be short-lived.

The second type of businessman is one who scores poorly on one dimension but highly on another (represented by squares in Figure 3.2). He would typically be either a businessman with very efficient production but low technology products (bottom left, Figure 3.2), or perhaps a boffin–entrepreneur who has no idea how to run a business but is blessed with a great technical expertise that enables him to produce a product for which there is strong though limited demand at virtually any price (top right, Figure 3.2), for example in the field of medical instrumentation. In both instances these entrepreneurs marginally qualify for the 'more successful' half of the matrix.

However, the most successful businessmen in Figure 3.2 are those who, as might be anticipated, develop *both* business acumen and

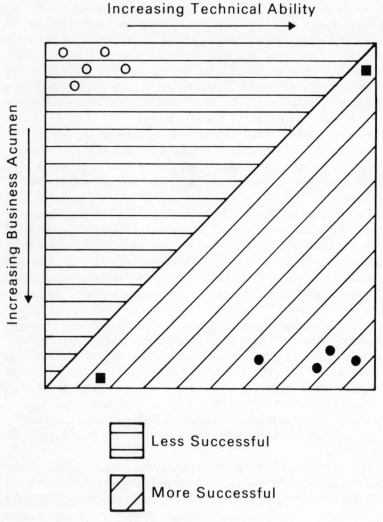

Fig. 3.2 An entrepreneurial matrix

technical ability (represented by circles in Figure 3.2). While the absent-minded professor-entrepreneur does exist, logic suggests that in most cases these (mainly white-collar) small firm entrepreneurs with technical proficiency, and almost certainly a higher education, are also more likely to be better able to develop the business acumen necessary to run a business as compared to blue-collar entrepreneurs with little technical knowledge (Johnson and Cathcart 1979). Indeed, Silicon Valley in California, famous for its small-firm growth, has been created by just such highly motivated, technically qualified entrepreneurs who have developed a flair for business (Cooper 1970). Perhaps Hewlett and Packard are the best examples of this phenomenon. None the less, while it is conceptually useful to talk in terms of these categories, in reality, the matrix in Figure 3.2 would be peppered with a broad spread of entrepreneurs *if* the matrix could be accurately calibrated. Clearly, the exact mix of individual entrepreneurial abilities are as unique as the human fingerprint and are thus virtually impossible to calibrate accurately in terms of Figure 3.2. However, evidence on high technology firms from subsequent chapters of this book confirms that it is the small firms with the entrepreneurs of the type indicated in the bottom right of Figure 3.2 that will, over a period of time, show the most vigorous growth and subsequently have the greatest impact on regional and national economies.

A further distinction is relevant to this consideration of entre-preneurial dynamism and its effect on small firm growth. The preceding arguments were careful to use the broad term 'activity' when discussing manufacturing in small firms. A frequent mis-conception when considering new firms is the belief that such enterprises are inevitably based on a new product idea. A cursory examination of the business behaviour of small firms indicates that this is not the case. Depending mainly on the mix of business acumen and technical ability embodied in the entrepreneur, the majority of new firms evolve in one of the following ways.

(a) Product-based firm. This category of firm is typified by the high technology firm with high levels of both business acumen and technical ability, and is represented by circles in Figure 3.2. Such a firm is founded predominantly on the basis of a new product idea. The product, because it is based on an in-house-developed specification and fills a relatively exclusive niche in the market, often achieves a high rate of sales and subsequent profit. The ability to produce a new product at the outset implies that in most cases subsequent products will follow in later years (model (a) in Figure 3.3). The combination of this continuing technical ability with the reinvestment of profits ploughed back into the business through internal research and development ensures the continued competitive edge and growth of the firm. Indeed, in high technology industry, product life cycles of

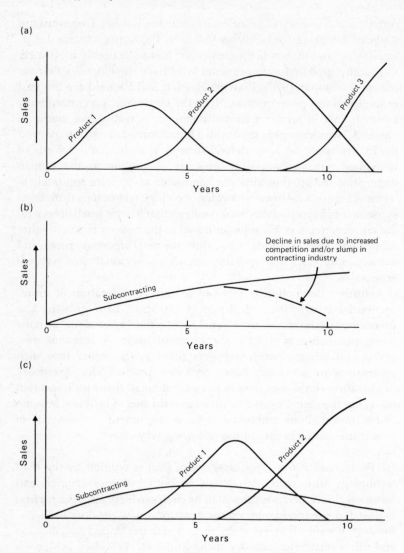

Fig. 3.3 A categorisation of new firm evolution

as little as five years are common, emphasising the need for constant research and development in order to avoid the limitations of a product with a short life expectancy. Such firms are the type of enterprise whose rapid growth puts life into any regional economy, as evidenced, for example, by Hewlett and Packard or Texas Instruments in the American context. These firms are the most sought after by regional development agencies attempting to stimulate indigenous industrial expansion.

(b) Subcontract firm. Model (b) in Figure 3.3 typifies a second category of firm founded on subcontracting. In most cases, the very competitive subcontracting environment means that profit margins are small. Consequently, growth is much slower than in the product-based firm. In such a subcontract firm, the entrepreneur frequently has little technical ability to develop his own product. Depending on the business acumen of the firm's management, subcontract firms would range between the upper and lower corners of the left side of Figure 3.2. Their owners are often ex-shopfloor workers with a good knowledge of process technologies but with very little idea of how machined parts might be transformed into a product. Because profits are generally lower than those achieved by the product-based firm, there is less capital available for research and development directed at developing an in-house product. Moreover the absence of a recognisable product in this subcontract category of firms may also blunt the interest of certain extenders of investment capital such as banks or venture capital organisations.

Apart from lower profit margins, a further major drawback to subcontracting is a frequent reliance for growth on the continued success of a single or limited number of customers. While dependence on a single major customer will be avoided by the subcontractor if possible, it is more difficult to achieve in areas where production is tailored to a particular subsection of an industry. For example Rabey (1977) found that British Northern region subcontractors to the shipbuilding industry were badly hit by the decline of the final-assembly stage in shipbuilding on the River Tyne. It is difficult to move production into another industrial sector in the short term. If a subcontractor machines large bearings for ships' engines, in a shipbuilding-dominated regional economy it is often impossible to find alternative work outside shipbuilding for which the firm has expertise. The development of a new product for such a firm might be even more traumatic. Apart from the absence of research and development capacity to develop new product ideas, new product development in a subcontract firm is further exacerbated by the frequent lack of a marketing infrastructure. This is caused by the nature of subcontracting in which, once a tender has been agreed, sales are assured for the duration of the contract without recourse to marketing. While firms seeking to escape from subcontracting might have a new product idea, they frequently have little conception of how successful it would be in the marketplace. Lack of marketing experience increases the risks involved in product development and stands as a barrier to a move from a subcontract to a product base.

(c) Subcontract–product transition. The third category of firm is the enterprise that does manage to make the transition from subcontracting to become product based. This category is a hybrid of the two previous types. Although the preceding discussion of subcontracting-based firms centred on the difficulties of transition from a subcontract to a product base, transition can take place if the entrepreneur has a strong personal motivation to progress and escape dependence on subcontracts. Subcontracting is always a risky business, not only because of the previously mentioned problems, but also because a subcontract firm with a high proportion of its sales to an important large customer may be placed in jeopardy by a sudden decision of the customer firm to perform the subcontracted task internally because the volume of business is becoming too large. Thus, most entrepreneurs are looking for long-term independence and security from such vagaries. A good product base offers a greater degree of internal control of sales and subsequent growth of the firm.

One method by which an entrepreneur with good business acumen but low technical ability can transform accumulated subcontract profits into a new product is by buying expertise. Given sufficient profits, resources can be ploughed back through the employment of full-time engineers in a small research and development department or, more modestly, through the services of an external consultant. The ensuing product can eventually be introduced to complement or replace the previous subcontract business of the firm (model (c) in Figure 3.3). However, such transitions are not common since technical ignorance about a new product idea on the part of the entrepreneur encountering this risk certainly inspires less confidence and enthusiasm than the case of the entrepreneur with the technical ability to create his own idea. For if the entrepreneur does not have intimate technical knowledge of a proposed product, and some idea of its subsequent sales, he is less likely to 'put his shirt' on any given project. None the less, a firm that does achieve such a transition is better placed to influence its own rate of expansion and embody a higher subsequent potential for significant employment growth in keeping with its newly acquired product-based status.

3.3 Industrial structural effects on the birth and growth of small firms

Given the 'ability' and 'functional' categorisations of firms discussed in the preceding section and their generalised relative growth potentials, what kinds of industrial structures are likely to generate these types of small firm? There is evidence to indicate that ageing industrial structures dominated by industries with mature products are poor 'seed beds' for new small firm formation (Johnson and Cathcart 1979; Storey 1982). Conversely, there is other evidence to suggest that local economies with a high proportion of newer high technology industries (particularly electronics) are not only good seed beds for the birth of small firms, but also fertile environments for their subsequent rapid growth (Cooper 1970; Little 1977; Bullock 1983). If these trends are brought to bear on the categorisation of firms hypothesised in the previous section, ageing industrial structures may be less likely to produce product-based firms of the type described in Section 3.2(a). New firms, when they occur in such economies, will be more likely to perform subcontract functions similar to those described in Section 3.2(b), since the majority of potential 'spin-off' entrepreneurs are predominantly blue-collar shopfloor workers with mainly process-oriented skills. This phenomenon has been noted in the past for the Scottish development region (Firn 1975; McDermott 1976).

While there will always be a few exceptions that prove the rule, there is a strong general rationale for these arguments. This rationale lies in the evolutionary stages of production through which most industries and their products pass. An explanation of the poorer entrepreneurial potential of mature industrial structures is aided by Figure 3.4. Here the typical development path of an industry is shown in a simplified form. The evolution of the actual products that comprise industries will always be to some degree at variance with this idealised pattern. For example, products of the electronics industry may reach maturity, become obsolete, and be superceded in less than

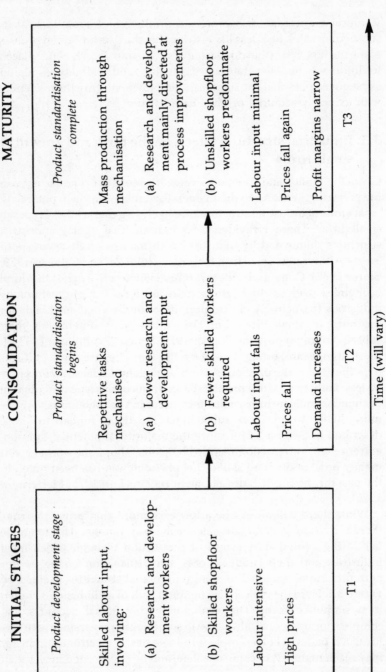

Fig. 3.4 A simplified model of industrial product development

ten years, while in the motor-vehicle industry the basic design of the family car (cosmetic revamps excepted) has changed little since the end of the Second World War.

Generally, there is a clear link between the maturity of industrial products and the applicability of mass production process techniques. If the products of an industry are high technology and have short life cycles, then there is little scope for the introduction of standardised process machinery that might free production from the constraints of the skilled labour inputs of both research and development and shopfloor workers (Segal 1962; Oakey 1981). Indeed, in many high technology industries, due to the very short product life cycles, products become obsolete and are replaced before they reach the 'consolidation' phase shown in Figure 3.4. However, the longer the product life cycle, the greater the scope there is for the mature phase of production and hence scope for the subsequent introduction of process machinery operated by unskilled shopfloor workers. The 'maturity' of an industrial product implies that it has a long standardisation phase which ensures sales *long after* radical specification change has ceased. These products predominate in the mature industries of depressed regions and their evolutions have frequently followed the pattern shown in Figure 3.4. Typically, such industries have a single, major, mature product such as textiles or ships.

However, the relevance of Figure 3.4 to the current discussion of the effects of industrial structure on the generation and growth potential of small manufacturing firms is that there will be a high incidence of new firms in the initial stages of an industry's development rather than in the mature phases (Rothwell and Zegveld 1982). This is due to two main factors as follows.

(a) An industry in its initial stages, when product development is rapid, will generate many sub-component niches for the entrepreneurs of new small supplier firms to fill. This is because constant modifications to large-firm products stimulate new product niches for supplier firms. Since mass production has not yet occurred in the large leading-customer firms of the sector, the small firms that emerge to feed them are able to cope with the frequently modest volume of production required. Also, because standardisation is unlikely at this stage, prices obtainable for sub-assemblies are relatively high, since the low-volume production runs, common in these small supplier firms at this juncture in the industry's evolution, make it unprofitable for large firms to enter such fragmented sub-assembly markets. While

such production niches are unlikely to ensure rapid growth immediately, they do offer new small firms a foothold in the emerging industry from which they may grow to generate subsequent products with greater sales potential.

(b) The need for skilled workers (*both* in research and development and on the shopfloor) to produce new products in the intial stages of an industry's development ensures the generation of a local reservoir of technically skilled personnel. These workers are most likely to 'spin off' locally and form product-based new firms which, as discussed earlier, have the greatest overall growth potential. In general, these are the new-firm founders with the greatest potential, most likely to generate significant innovations and hence, to offer real prospects for employment growth.

These two factors potently interact to the benefit of both large and small firms in areas with a predominance of new industries. The multifaceted high technology electronics sector is a perfect 'real world' example of such factors at work. Indeed, the electronics industry displays an amoeba-like characteristic in that, no sooner has one product division of the industry become mature or obsolete than another product development breaks away from the 'mother' industry to take its place. The technological development from the electronic relay, through integrated circuits, to the current microprocessor is a good example of this regenerative phenomenon.

However, other industries appear to display all the characteristics of product maturity and stagnation outlined in the third stage of Figure 3.4. These are the industries, common in the depressed regions of Western economies, in which Third World countries have captured an increasing market share since the Second World War (for example textiles, steel and shipbuilding). In such forms of production, since there is little scope to revive prices through improvements in product performance, process costs become all important. Such an austere and exacting production environment is not conducive to the birth of small firms through spin-off. Few product niches remain for potential entrepreneurs to exploit and the inevitably small production levels implicit in new small firms are not suited to industries where profit margins are small and mass production abounds. The propensity of small firms to produce high-cost output in small amounts is not consistent with the general requirements of cost-minimising large production scales. Moreover, unlike embryonic industries, the standardisation of production in these large-firm-dominated industries ensures that research and development and the number of skilled shopfloor workers will be minimal. Thus, most new entrepreneurs emanating from these mature industries will be blue-collar workers. Moreover, it is worth restating that apart from the generally poor business acumen of these new entrepreneurs, such workers are more likely to be proficient in shopfloor process skills and thus show a propensity for the type of subcontracting activity shown in model (b) of Figure 3.3, with all its attendant disadvantages.

If these arguments on the effect of industrial structure are projected into a regional context, the implications are clearly unfavourable for the economies of the depressed development regions. Many of the staple industries of these regions have reached the high level of maturity shown in the final stages of Figure 3.4. Consequently, the potential for new firm formation based on new products with growth potential is much reduced and further exacerbated by the development regions' limited mix of skills, which are heavily blue-collar oriented and research and development deficient. The problem is compounded, and regional disparities increased, by the assertion that the reverse is true of the prosperous regions, as evidenced by the industrial growth in the Thames and Silicon Valley areas of Britain and the United States respectively. Such regions have a disproportionate share of dynamic new industries and, consequently, of skilled production and research and development workers. Seen in this light, the inherited industrial structure of a region plays a large role in determining the pace and direction of current and future industrial growth. If the industrial structure was diversified and expanding in the past, this trend will boost current performance; if stagnation was prevalent, this trend will also be reflected in current conditions.

4 Agglomeration, innovation and high technology industrial production

4.1 The principle of local agglomeration economies

Even to an inexpert eye, an aerial view of a great industrial complex will, through its very physical interconnectedness, suggest that there is some totality to such a development that is greater than the sum of its parts. Moreover, at ground level, conversations with individual industrialists within the complex will readily isolate a heterogeneous range of factors that yield significant advantages of location to their firms. Such conversations give organisational meaning to the local physical linkages that produce agglomeration economies from the close juxtaposition of various complementary forms and stages of industrial production. These relationships are apparent in many diverse forms of manufacture, for example the steel-based industries of the Ruhr or the textile areas of the North West of England, where every stage of production, from processing the raw iron ore or cotton to the manufacture of finished or semi-finished products, is functionally linked together within integrated large corporations, a mass of smaller firms, or a combination of both.

From the time that the isolated cottage industries of the eighteenth century began congregating to form the great industrial areas of the Victorian era, it has been apparent to economists that such focused industrial activity, or agglomerations, were generating spatially discrete advantages over other dispersed production locations. This point was not lost on early location theorists who readily sensed the pervasive advantages of the agglomeration as a key variable in theoretical equations designed to determine the optimal production location.

At the turn of the century, theorists in many emerging social sciences were preoccupied with the task of adapting precise laws from mathematics and the physical sciences to explain quasi-social phenomena. Alfred Weber's contribution to location theory (Weber 1909) was an example of this type of work. However, although Weber acknowledged the production advantages of spatially concentrated industry, the inclusion of this agglomeration variable in his theory of location caused considerable problems of definition and measurement. Notwithstanding the overall theoretical simplicity of his 'cost-minimising' model, transport and material costs could be accurately quantified and therefore incorporated into his location 'equation' as steps on the road to determining the 'least cost' location for production in relation, in the case of iron production, to a hypothetical single market and two raw material sources. But although other factors such as political considerations were not included in his location model, perhaps Weber's greatest folly was to oversimplify the effects of agglomeration economies (Lloyd and Dicken 1977). For Weber, agglomeration was a term limited to considerations of production scale advantages in the plant, or clustered plants in a local area, together with an inadequate treatment of 'accidental' agglomeration economies deriving from urban or transport mode locations (Riley 1973). In fact, agglomeration is a term that describes a heterogeneous collection of both small and large economies stemming from the close juxtaposition of industrial firms, frequently located in urban areas.

The unrealistic partial consideration of agglomeration economies was perpetuated by the later work of August Lösch, who founded the 'market area' or 'maximum profits' school of industrial location theory. The concept of agglomeration was also central to his work in that the motive of maximum sales was used in determining the ideal maximum profit location. According to Lösch, the ideal location would be one at which maximum sales could be achieved. While this principle was relatively clear when applied to a single plant and a single market area, the model became very complicated and confusing when an attempt was made to assess the agglomerative advantages of many plants in many overlapping market areas. None the less, it is clear that agglomeration advantage is implicit in the resultant concentration of sales-maximising production into 'city-rich' sectors (Lösch 1954). However, the important point of relevance to the current discussion is the partial consideration by both writers of the pervasive concept of agglomeration, and the emphasis placed by these early theorists on one agglomerative advantage at the expense of others. In reality, the advantages of location in an industrial agglomeration, judged at a bare minimum, accrue from *both* material input and sales output linkages. Indeed, more recent writers on linkages have found that other subtle economies are also central to an explanation of the total effect of agglomeration on industrial location (Wood 1969; Taylor 1973; Gilmour 1974).

The study of linkages has a long and distinguished history within industrial location research. Early post-war work sought to emphasise the importance of material input and output linkages and the subsequent agglomeration advantages that ensued from particular industries being in specific industrial locations. Studies of the gun and jewellery quarter of Birmingham (Wise 1949) and the instrument (Martin 1966) and furniture (Hall 1962) manufacturing industries of inner London all examined the purchasing and sales relationships of producers in these small-firm-dominated 'disintegrated' industries. However, explicit flows of material inputs and outputs between manufacturers were implicit evidence of other behavioural advantages of close and frequent personal contacts. For every flow of materials, there was a parallel flow of technical or commercial information. There could be little doubt that, given the methods and scales of production operated in these industries, agglomeration advantage through material *and* information linkages was a strong force towards industrial location inertia.

However, beginning in the 1950s, and intensifying in subsequent years, there was a sharp decline in many agglomerated urban industrial concentrations. For example, the industries examined in the earlier studies began to decline, for several reasons. A tendency towards larger scales of production within large firms in certain industries, cheaper foreign competition, and the increased costs of an urban location through increased rates, land costs and transport congestion tended to diminish and disperse these archetypal agglomeration-based industries that were best-suited to the industrial conditions of the Victorian era.

The growth industries of the 1960s were in sectors typified by electronic consumer goods. These industries were frequently termed 'footloose' because they were not spatially constrained to specific locations by labour, customers or suppliers. With a continuing decline in the cost of transport, and the generally high value per unit of weight of both inputs and outputs to the production process, it was argued that almost any location with an abundant and preferably cheap labour supply would suffice (Fulton and Hoch 1959). On the basis of these economic arguments, and with the stimulus provided by 'carrot' and 'stick' government incentives in favour of relocation, the mobile-industry boom of the 1960s saw the diversion of much industrial expansion from the West Midlands and South East of England to the development areas (Keeble 1976). At the centre of this mobility concept was the assumed reduced importance of local linkages to industrial location. Feedback from detailed research in relocated and new branch plants in development regions presented a mixed picture. In some instances the loss of close material and information links with key suppliers and customers caused serious difficulties (Luttrell 1962; Townroe 1971). Moreover, in assessing the benefits of the previous location in, say, London, executives were not restricting their comments to the importance of material input or output linkages in a respectively Weberian or Löschian manner. For example, a common problem in the months following relocation was that of the acquisition of suitably skilled or trainable workers (Townroe 1975; Oakey 1979b). In many cases these early post-move problems were 'ironed out' within two or three years of the move. However, the significant closure rates recorded during this period bear witness to the fact that many of these industries were not as 'footloose' as originally anticipated (Sant 1975; Smith 1979).

In summarising the nature of writings on industrial linkage during the 1960s and early 1970s, two important features are distinguishable from the contributions of academics. First, there was an intense burst of research into the locational impact of linkages on the mobility of industrial firms (Brittan 1969; Karaska 1969; Townroe 1971; Moseley and Townroe 1973). Unlike the earlier studies of the impact of linkages on individual industries in established locations, these studies were mainly multi-sectoral and focused on the problems of local linkage adjustment in relocated firms (Luttrell 1962; Townroe 1971). Second, there was a lengthy theoretical debate on the definition of agglomeration economies as represented by industrial linkages. Wood (1969) argued, on the basis of accumulating evidence from field surveys, that the advantages of an agglomeration were insufficiently described by either material input or output linkages. He introduced the concept of 'information linkages' in industrial interactions as a component of the *total* attractive and retentive powers of an agglomeration. It should be re-emphasised that both technical and business information was an implicit aspect of the older agglomeration economies noted by Wise, Martin, and Hall for industries in Birmingham and London. None the less, Wood's attempt to widen agglomeration advantage provoked a response from Smith who argued that the definition of agglomeration economies should be strictly limited to material flows (Smith 1970). Other contributions to the debate in *Area* developed the arguments further (Taylor 1970; Bater and Walker 1970).

However, apart from the debate on whether material or information linkages constituted agglomeration economies, Gilmour (1974) further argued that, for labour-intensive industries, an attractive labour force might be the *most important* benefit of an agglomeration. In retrospect, the extended debate appears to owe more to Smith's attempts to 'patch up' the damage done to Weber's original theory by the growing complexity and sophistication of industrial production, and to erase initially inherent defects in the model, rather than accurately define agglomeration economies in a modern context. With regard to the developing argument of this chapter, it is sufficient to note that the influence of information and labour advantages were added to those of materials by many writers when considering the constituents of industrial agglomeration advantage.

Generally, the end-product of all these investigations into the importance of linkages to agglomeration economies in modern industry tended to confirm the reduced importance of customer and supplier links to the location of the expanding industries of the 1960s (Karaska 1969; Moseley and Townroe 1973). The problems caused by information and labour disruption were not in most cases considered sufficient to cause long-term problems (Townroe 1971, 1975), with the significant exception of industries with highly skilled shopfloor workers where continuing labour problems were common (Oakey 1979b). These rather negative, dismissive findings on the location importance of linkages, together with the onset of the 1973 oil crisis and the subsequent arrest of industrial growth and mobility in most sectors, tended to sap interest in the importance of linkages and agglomeration economies to the mobility of industry.

4.2 High technology industry and the principle of agglomeration

Notwithstanding the distinguished history of the agglomeration principle in acting as a basis for diverse theories of industrial location, the following discussion considers whether this long-standing concept remains obsolete in the high technology growth industries of the closing decades of the twentieth century. The 1970s were notable for the accelerated decline of industries once staple to Western industrialised economies. One by one the traditional industries of iron and steel, shipbuilding, motor vehicles and consumer electronics have been diminished by a flood of high-quality, low-price manufactured goods from Third World or quasi-Third World countries. In such circumstances, the prosperous industries of Europe and the United States are now frequently defined as those that are declining *less* than the national average. This industrial decline, and the subsequent decline in demand for additional production floorspace, must explain much of the concomitant decline in the volume of mobile industry. The motor-vehicle sector is a perfect British example of this phenomenon. During the 1960s, relocation of expanding motor-vehicle production was a symbol of the achievements of British mobile-industry policy (Goodman and Samuel 1966). Development region plants in Scotland, the North West and Wales brought much-needed employment to the unemployment blackspots of these regions. The subsequent decline of the British motor-vehicle industry branch plants in particular, and of production relocated to development regions in general, has highlighted the extreme way in which the mobile-industry palliative has collapsed, in terms of both the current dearth of relocatable production and the poor subsequent employment records of many firms relocated to development regions in previous 'boom' years (Smith 1979).

But set against this general picture of British industrial decline since 1970, there have been some success stories, notably in the electronics industry where rapid technological advances have extended the scope for new manufacturing at the high technology end of the market. Although not totally compensating for the decline in other sectors, high technology electronics-based industries do offer some salvation to national and regional economies. Consequently, much current discussion between experts in the fields of industrial location and regional development is centred on the potential of these new high technology forms of production for the alleviation of industrial decline in depressed industrial regions, either through indigenous growth (Ewers and Wettmann 1980), or through the acquisition of expansion-stimulated, mobile, high technology production from prosperous areas in which it is currently prevalent (Oakey 1983).

However, it would be foolish to assume, without prior consideration, that such expanding high technology forms of production will not be sensitive to local agglomeration economies merely because the industries that fuelled the previous mobile-industry boom were generally insensitive to such stimuli. If the relocation of expanding high technology production from prosperous to depressed areas is to be proposed, the mobility of such production must be seriously considered. The following discussion of the mobility of high technology industry will rely heavily on the theory of the product life cycle in an attempt to link the consideration of the contemporary location requirements of innovative high technology industry to the general concept of agglomeration discussed extensively in the preceding section (Vernon 1966).

It can be generally asserted that the duration and shape of sales curves achieved in the course of a product life cycle will greatly depend on the industry in which the product is developed. If the example of the consumer electronics industry is taken, then the thirty-year product cycle of model (a) in Figure 4.1 might be typical for an electronic consumer durable, albeit with appropriate facelifts (for example vacuum cleaners). Several implications flow from this form of product life cycle when it is related to development regions and their experiences as hosts for much mobile industry since the Second World War. The evolution of the product life cycle implies several production phases. First, in the early stages (stage 1), there will be many teething problems surrounding the product's initial construction on the shopfloor. This development stage will require substantial inputs of skilled workers, both in research and development and on the shopfloor. In these early stages, the manufacture of the product is most efficiently located in an industrial agglomeration in order that links might be maintained with a headquarters or a research and development centre commonly located in such an area (Goddard and Smith 1978; Buswell and Lewis 1970), and in order that recurrent changes in labour and material inputs may be met by a rich and flexible local resource environment. During the period of maximum sales (stage 2) the production methods related to the product are perfected and the level of skilled labour input, both in research and development and on the shopfloor, is reduced. In addition the standardisation of the product enables material inputs to be standardised and purchased in bulk on a regular basis. Thus, during stage 3, the product may be transferred to another peripheral location where development grants and cheaper unskilled (probably female) labour may be obtained to prolong the product's life through the reduction of production costs. Production continues in this location until, eventually, obsolescence is reached at the end of the 30-year cycle (model (a) in Figure 4.1).

This typical cycle has much relevance to the previous discussion of mobile industry where, in particular, the term 'footloose' could be unequivocally applied only to the third stage of model (a) in Figure 4.1. The standardised electrical and mechanical engineering manufacture transferred to the development areas in the 1960s, the keenness of firms in this type of manufacture to obtain government grants towards the purchase of capital equipment, and the propensity

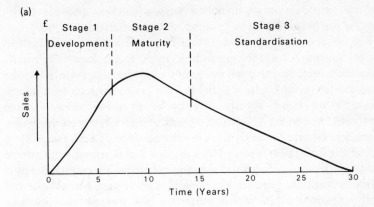

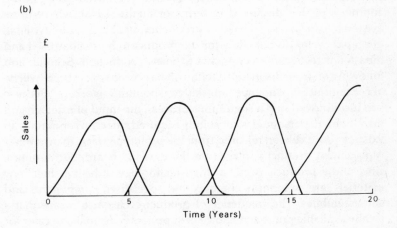

Fig. 4.1 Long and short duration product life cycles

for the employment of unskilled female workers (Keeble 1976) suggest the 'stage 3' nature of much of this mobile production.

But the above phenomena do not currently apply to product life cycles in much of the expanding high technology electronics industries. Again using the concept of the product life cycle, it is clear from model (b) in Figure 4.1 that the life cycles for high technology products are much shorter. Life cycles in these industries are commonly five to ten years. The major difference between the cycles of models (a) and (b) in figure 4.1 is the absence of a tail (stage 3) in model (b). This tail is replaced by a decline that is virtually the mirror image of the incline. Sales and subsequent profits are maintained by a series of multiple product life cycles (often arising from the initial establishment of new firms) that ensures overall viability and growth.

Several locationally important factors result from this excited rate of product innovation which create location behaviour that is the antithesis of that discussed in terms of Figure 4.1(a), where there existed a standardised phase of production enabling mobility. First, the rapid cycles do not allow for the product to be standardised and thus for it to become conducive to mass production. Second, and following on from the inapplicability of mass production, there will be a continuing dependence on skilled shopfloor workers. Because product innovation is a continuous process, the input of research and development personnel is constantly required since, no sooner has an existing product been released from the prototype stage, than another replacement product is placed on the drawing board. Furthermore, since these products occur in high technology industry, their 'raw materials' are commonly complex and specialised components and sub-assemblies. The frequency of product cycles also means that a locally available range and choice of suppliers is desirable to cater for early modifications in design, as was the case in stage 1 of model (a) in Figure 4.1. Thus, in a sense, the products of model (b) in Figure 4.1 never evolve past the development stage of model (a), certainly not in terms of becoming indifferent to the advantages of local agglomeration economies common in the standardisation phase of model (a).

These observations have three major ramifications of significance to the current consideration of the effect of agglomeration on high technology firms and their subsequent regional mobility. First, the high technology product life cycles simplified in model (b) of Figure 4.1 suggest that there is no standardised or 'footloose' stage to product life cycles in those high technology industries that might be relocated to development regions. This phenomenon must in part explain the general inability of the high technology firms that have been expanding during the 1970s to produce contributions to mobile industry through spin-off expansions in depressed regions.

Second, the previous arguments on the significance of 'traditional' agglomeration economies are revived by these comments on product life cycles. It is likely that, in the light of the preceding arguments, the concentrations of high technology industry in the Silicon and Thames valleys, or on Route 128 outside Boston, are agglomerations in the 'traditional' sense, similar to the previously described urban-agglomerated industries of the past. Certainly, the hypothesised continued emphasis on local skilled labour and on both local material and information linkages is very similar to the agglomeration advantages common in the traditional agglomeration. The suggestion that modern high technology industry is heavily agglomerated will be closely considered in the following, empirically based chapters.

Third, there is obvious scope for small firm formation in any industry experiencing rapid product change. Because large firms in a high technology industrial agglomeration are in a continuous process of product change, standardisation is difficult. Hence, as discussed in Chapter 3, in any area of production where standardisation is difficult and prices subsequently high, there are niches for small firm operation. Indeed, the linkage economies of high technology industrial agglomerations are mainly based on the specialist support services of small firms which can cater flexibly for the precise service and component needs of their larger patrons. This large firm–small firm symbiosis is very evident in Silicon Valley, where firms such as Fairchild, Hewlett and Packard, and Varian provide market scope for the, now famous, small firm growth apparent in the area.

Hence, because these modern high technology agglomerations have great potential for internal growth, they have a tendency to reinforce their own competitive advantage at the expense of other areas where agglomeration advantages do not exist. Such was always the tendency in older agglomerations and, in a sense, these economies define an agglomeration. While there is some evidence that diseconomies do increase due to competition for labour and land at the heart of agglomerations, these negative factors are predominantly endured in order to reap the overwhelming benefits available (Alonso 1971). In any case, any development region benefit from relocated production resulting from the diseconomies of such congestion would be low, since the high costs of relocation would tend to encourage development on the fringe of the agglomeration rather than in some far-off region with few agglomeration economies. And for those few high technology products that do generate a high enough volume and a product cycle long enough to justify mass production (for example random access memories and computer toys), the most common practice today is to overlook the development areas of Western developed economies and relocate production to South-East Asia in order to take advantage of relaxed industrial practices (for example in pollution control) and cheap non-unionised labour.

4.3   High technology agglomerations and regional small firm innovation

The preceding arguments have generally indicated that high techno-logy industries spontaneously agglomerate to obtain linkage, labour and information advantages. These are the industries with the greatest recent growth in modern Western economies, where sophisticated high technology, embodied in sophisticated product specifications, frequently protects European and American producers from Third World competition. However, given current economic conditions there is little likelihood, in view of the preceding arguments on the lack of product life cycles with long phases of standardisation, that much of this high technology production will spontaneously relocate to development regions without the spur of government policies. Moreover, it is clear from the evidence of Chapter 2 that the development of high technology production is unlikely to ensue from indigenous development region innovation. In the development regions, the overall low level of product innovation in independent small firms, and the significant influx of innovations from outside the region into branch plants, to some extent confirm this poor indigenous innovation performance. It should be re-emphasised that the 'new' product innovations in these branch plants may only be new to the factory concerned and not new in an absolute sense. In fact, the 'new' product innovation in such an establishment might have been transferred from another production site and be at the 'standardisa-tion' stage of model (a) in Figure 4.1.

This chapter's largely hypothetical assertions on the importance of agglomeration advantage to successful innovation in high technology industry supports the major rationale of the research on which this book is based. The material and information linkages and labour advantages, depicted in this chapter as strong forces of agglomeration, are largely synonymous with many of the local innovation resources examined in subsequent chapters. In this sense, the concept of 'agglomeration' may be used interchangeably, and to the same effect, with the term 'local resource environment' to describe in aggregate the individual local resource advantages of an area that assist product innovation and growth in industrial firms. Hence, it is clear that the principle of agglomeration economies is central to the main argument of Chapter 2 on the variable effects of local regional resource environments on small firm innovation and growth. Thus, the hypothesised causal link between the quality of local resource environments and innovation levels in small firms might be termed an agglomeration effect.

Consequently, the following analysis, under specific chapter headings, of the impact of individual resource factors on innovation in small high technology firms may be seen as a wide-ranging attempt to test the impact of individual agglomeration components on the innovation and growth of small firms in sharply contrasted regions. Hence, the aggregation of these individual results will provide a basis for the assessment of the total effect of agglomeration advantage on innovation.

Later chapters will also test the assertion that high technology industry is a form of production that strongly benefits from agglomeration economies and is therefore strongly subject to forces of inertia. Such a constraining effect would have important implications for any policies aimed at the regional dispersion of high technology production from agglomerations in order to give development regions a share of this growing area of manufacturing industry. The investigations of Chapters 6 to 9 will provide empirical evidence on the extent to which high technology industry is agglomerated, and this will be discussed in the conclusions of Chapter 10.

5   Methodological and environmental contexts

Thus far, discussion has mainly focused on conceptual justifications for the following detailed study of variations in local environmental resource factors and their impact on innovation in small high technology firms. However, in advance of chapters that examine the specific importance of individual innovation-influencing factors, some contextual account should be made of the research methodology, the character and industrial performance of the individual study regions, and their overall innovativeness. The general measures in this chapter of such innovation outputs as product and process improvements will provide a hard statistical background on relative regional innovation performances. These data will facilitate a valuable contextual insight in advance of the detailed analysis of the individual resource factors that combine to produce these innovation outputs.

5.1   The methodological approach

5.1.1 The survey design

An initial decision on methodology concerned the method of investigation. Previous experience with the use of secondary data on innovation (Oakey et al. 1980) and with the combined use of postal and interview questionnaires (Oakey et al. 1982) strongly suggested that, a higher unit cost of questionnaire completion notwithstanding, the interview questionnaire method was far superior to any other available means of data collection. Moreover, this method seemed particularly appropriate for the study of small firms since it permits the acquisition of valuable anecdotal information with which to set the hard quantitative data in a broader socio-economic context. For, in small firms, the actual behaviour of executives is only the surface manifestation of more personal motivations that are directly related to the personal philosophies of the firm's management (Senker 1979).

5.1.2 Choice of regions

Clearly, the successful measurement of differences in innovation-influencing regional resource environments depended heavily on the selection of appropriate survey regions. By considering small independent firms of similar size, status and industrial type, the regional environment was allowed to vary sharply in the expectation that regional variations in innovation levels would largely be a reflection of the effect of diverse *external* regional environments, and not due to *internal* differences in firm characteristics. The choice of study regions reflected this desire for the greatest possible environmental diversity. The potentially least innovative of the regions chosen was the British development region of Scotland, with a history of industrial decline in basic heavy engineering sectors since the turn of the century, particularly in the post-war period. The South East of England was the second region included in the study. This region has consistently been the most successful of all the British standard planning regions. The environment of the South East, with the advantage of a non-federal national government system based in London, the benefit of a concentrated, corporate, decision-making infrastructure (Parsons 1972; Goddard and Smith 1978), and a dominant share of institutionalised public and private research and development (Buswell and Lewis 1970), has recorded significantly higher levels of industrial innovation when compared with the other regions of Britain (Oakey *et al.* 1980, 1982).

However, a restriction of the study to British regional environments did not wholly meet the principal objective of study region selection, which was that *maximum* environmental diversity should be achieved. While initial investigations in the British study regions yielded valuable evidence on innovation-influencing variations in the differing environments of Scotland and South-East England, in other instances there was evidence of the pervasive influence of British centralised unitary government with its conforming effect on the economic and political environments of firms in Scotland and the South East, regional incentives notwithstanding. Moreover, the oligopolistic behaviour of the 'big four' British national banks through the similar range of services and interest rates offered to customers tends to reduce the spatial impact of British regional financial environments on innovation, since the attitude of British national banks in Glasgow will not differ markedly from those in London, particularly if the banks concerned are of the same group. (However, these comments do not deny the significant marginal impact of the Scottish banks, which is considered in later chapters.) Also, with only a few notable exceptions, links between British universities and industry have been noted to be uniformly poor throughout Britain (Oakey 1979c; Oakey *et al.* 1982). In general, there is a pervasive 'ivory tower' attitude in many quarters when university links with industry are considered, which must, in part, reflect national attitudes towards the status of manufacturing industry and its perceived contribution to the fabric of British society.

The decision to include a particularly innovative region from outside Britain increased the diversity of the study through the introduction of a different political and economic system, together with the subtle and pervasive influence of a fresh set of cultural approaches to the role of industry in society. The San Francisco Bay area of California was chosen for this international extension of the research because of its general reputation for rapid growth in the high technology industries of this study (Little 1977; Bullock 1983), and in particular for its high rate of small firm birth and growth based on an aggressive business ethic.

The decision to extend the fieldwork to the Bay area brought several specific advantages. First, the federal political system of the United States contrasts with the British unitary system. Second, there was some evidence to suggest that the high technology industrial agglomerations of Silicon Valley and Boston's Route 128 are closely linked with universities, both in terms of technical information flows and in entrepreneurial spin-offs (Deutermann 1966; Gibson 1970). Clearly, if various advantages accrue to small firms in California from university links, the exposure of these benefits might contrast significantly with the experiences of potentially less fortunate firms in the British regions. Third, there has been much recent, largely unsubstantiated, comment on the impact of venture capital on small firm innovation and growth in Silicon Valley. Since the acquisition of investment capital is a key prerequisite of any meaningful programme of innovation through research and development, the relative availability of venture capital might be of major importance in explaining variations in the overall performance of small high technology firms in different regions.

5.1.3 *Choice of industries*

A further decision concerned the choice of industrial sectors for the study. It was considered necessary that the survey industries should be high technology and highly innovative in order that they might put maximum stress on the resources of their local economies with a view to exposing any deficiencies. For example, it would in most cases be inappropriate to question the owner of a soft toy firm on his links with universities and their effect on his level of product innovation. Given that accumulated expertise existed from previous research on the instruments and electronics industries (Oakey 1981; Oakey *et al.* 1982), scientific and industrial instruments (MLH 354) and radio, radar and electronic capital goods (MLH 364) were again chosen as the subjects for this study. These industries are known to be highly innovative (Thwaites 1978; Oakey *et al.* 1982), and combine well as examples of high technology industry since, in practice, there is often considerable technological overlap between the output of the two industries, particularly at the small firm level.

In order that consistency might be maintained in this international study, the closest possible match was made between the production activities of the British and American survey participants. With the aid of a broad experience of individual products in the chosen British industries, the two British MLH categories were matched with groups of American four-digit industrial categories, individual examples of which virtually corresponded to discrete products. For a precise description of the British MLH and American SIC categories used in this study, see Appendix 1.

5.1.4 *Sample selection*

In a trade-off between the desire for statistically significant numbers and an acceptable level of cost, it was decided to aim for a survey sample of 60 firms per study region, divided equally between the two study industries. This size of sample would allow for a degree of disaggregation within regions during analysis, while retaining a measure of statistical significance. Universes of small, independent, single-site firms of less than 200 employees in the chosen industries were compiled in the British and American study regions. The original objective of the research was to establish large universes of small firms and select survey participants on the basis of stratified random sampling according to employment size cohorts, with replacement, in order that the size structure of the sample should reflect the original universe. Three main directory sources were used for this task. The British universes were compiled from the David Rayner instrument and electronic directories and the 1976 Census of Production business directory. The American data were derived from the California manufacturer's directory for 1982, which includes approximately 70 per cent of all manufacturing plants in the State of California.

While numbers were sufficient to allow random sampling in the

South East of England and the San Francisco Bay area of California, it was not possible to find 60 firms in Scotland with the required characteristics. Even after the addition of firms provided by the Scottish Development Agency, the final number of enterprises interviewed in Scotland was 54, 6 short of the target number. However, the fact that the 54 firms interviewed were virtually all of the 56 firms identified means that in the Scottish region the sample firms are almost the total population of identified firms. The different derivation of the Scottish survey firms should not, however, introduce any particular bias to the total survey data and is, in any case, an unavoidable ramification of the paucity of high technology industries in the British development regions. Indeed, it would not have been possible to conduct this study in the other development regions of Wales and the North of England due to a dearth of surveyable firms in these areas.

The overall response rate for Britain was a high 87 per cent, with a slightly lower level of 76 per cent in the San Francisco Bay area of California. The interviews began in England and Scotland during the summer of 1981 and were completed in California in the summer of 1982. The final survey total from the three study regions was 174 small firms. Further insight and consistency has been given to the research by the author's personal execution of all the interviews.

5.2 The contribution of the study industries to regional industrial employment

As a first step in an attempt to give a picture of the survey industries and physical environments present in the study regions, the following discussion will be based on aggregate published data. Clearly, since the methods of classifying industries differ between Britain and the United States, precise international comparisons are not possible. However, the presentation of the following data at the order level for Britain and the two-digit SIC number level for the United States represents approximately similar types and levels of industrial categorisation. Hence, it is broadly possible to equate the American instruments and electronic component sectors (SICs 3800 and 3600 respectively) with their equivalent British instrument and electronic engineering sectors at order level (Orders VIII and IX respectively). The actual American four-digit SICs and British MLHs chosen for this study are contained within these broader definitions, and are detailed in Appendix 1.

The objective in presenting these data is to give the reader a contextual picture of the strength of the instrument and electronic engineering industries in the study regions. It is essential that comparisons be made at a general level, not only because of the differences in sectoral definition between Britain and the United States, but also because the international regions, although approximately compatible in geographical area at least, are part of widely different national economic systems in terms of both economic and physical scale. For example, the San Francisco Bay area of California, similar in size to the South-Eastern planning region of England, is part of a state that is larger than the whole of Britain, which in turn is part of the United States, a country which dwarfs the free trade area of the European Economic Community. However, it is also true that California, due to its physical isolation from the central and eastern parts of the United States, acts to some extent as an independent industrial region, certainly in terms of its self-sufficiency in many of the resource inputs to the West Coast manufacturing economy. Hence, for the purposes of the following data presentation, the British planning region data are expressed in terms of total British industry, while the Bay area data are expressed in terms of total Californian industry. Clearly, this approach is not ideal, but it does provide a useful contextual picture, both of the overall employment contribution of the industries to the British and Californian total economies, and of their comparative growth performances over the 10-year period 1968–78.

One further comment on the following data is essential. The San Francisco Bay area of California is a reasonably coherent and contiguous socio-economic area stretching from Monterey in the south to Santa Rosa and Healdsberg in the north, and from the Pacific Ocean in the west to the foothills of the Sierra Mountains in the east (Figure 5.3). However, from the government statistical viewpoint, there is no meso-level statistical unit between the state and county levels. Hence the following data for the Bay area 'region' is an amalgam of county data selected on the basis of the greater San Francisco Bay area as defined by this study (for details of the individual counties see Table 5.2). The counties listed in Table 5.2 provided the spatial basis for the sampling universe from which the final survey firms were drawn (Figure 5.3).

5.2.1 *Scotland*

Although Scotland is a physically large region by British standards, most of its manufacturing activity is contained within the central

Lowlands around and between the twin foci of Edinburgh and Glasgow. It will be apparent from Figure 5.1 that, apart from a few firms further north and south of the central belt in the Highlands and southern Uplands, the majority of small high technology firms in this study conform to the Edinburgh–Glasgow corridor of the Lowlands manufacturing belt. Scotland is fortunate, compared with other British development regions, in having built up significant instrument and electronic engineering sectors since the Second World War, based on the attraction of large British and American branch plants (for example Ferranti, Hewlett and Packard, and IBM).

Hence it is not surprising that Table 5.1, which shows employment in the electronic engineering (Order IX) and instrument engineering (Order VIII) sectors of Scotland, records a substantial level of employment for the base year of 1968. This is particularly impressive given the development region status of Scotland. Over the decade 1968–78, during a period of general industrial depression which intensified after the oil crisis of 1973, there have been modest increases in the employment levels of both these industrial orders in Scotland. Indeed, measured in employment terms, electronic engineering was the only industrial order-heading in the top five Scottish industrial sectors marginally to increase its workforce during the decade. In comparison, employment in textiles fell from 83,000 to 56,000 workers during the same period. Set against this typical development region sectoral decline, and the overall national

Table 5.1   Employment in instrument engineering (Order VIII) and electronic engineering (Order IX) for Britain, Scotland and South-East England, 1968–1978

	Order VIII				*Order IX*			
	1968	*1978*			*1968*	*1978*		
	Number	*Number*	*Job change*	*% change*	*Number*	*Number*	*Job change*	*% change*
Scotland Employment	15,800	17,600	+1,800	+11.4	41,000	47,800	+6,800	+16.6
South East Employment	90,100	74,200	−15,900	−17.6	279,500	249,000	−30,500	−10.9
Britain Total	169,100	154,400	−14,700	− 8.7	743,100	677,100	−66,000	− 8.9

Source: Department of Industry (1968, 1978).

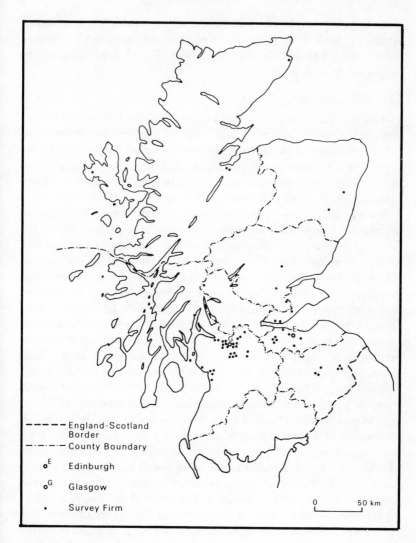

Fig. 5.1 Scotland

manufacturing employment decline evident in Table 5.1, this modest Scottish growth in both the electronic and the instrument sectors is encouraging, particularly since these sectors have good medium- to long-term growth potential.

5.2.2 The South East of England

The overall visual impression in Figure 5.2 of the location of survey firms in the South East of England reveals a predictable pattern, since the distribution of these high technology small firms reflects the general spread of industry outwards from a Greater London location in the immediate pre-war and post-war periods into the Home Counties north, west and south of the capital city (Oakey 1981).

Statistical evidence from the South East of England creates an impression much different from the picture of growth noted for Scotland in both the electronic and the instrument engineering sectors. The loss during the decade of 30,500 and 15,900 jobs in electronic and instrument engineering respectively is a substantial employment decline. A proportion of this fall in electronics is possibly due to improvements in production methods in the lower technology areas of this sector, where the substitution of capital equipment for labour has taken place. This overall sectoral and regional decline in employment might well mask employment increases in the high technology production niches known to be growing rapidly in the Thames Valley sub-regional economy. However, the particularly sharp decline in employment in the South East must be due, at least in part, to production relocated away from the South East. In particular, the exodus of manufacturing firms from Greater London since 1968 must be a strong contributory factor in this employment decline. Firms vacating factories in London or the inner South East frequently leave the South East altogether in favour of adjacent planning region locations (Keeble 1978). However, the national decline in employment in both sectors (Table 5.1) suggests that this regional decline may also reflect an overall national decline. None the less, when the three study regions are compared, it remains true that the South-Eastern planning region contains by far the largest number of employees in both the instrument and engineering sectors. Indeed, electronic engineering in the South East continued to be the major employer of the region in 1978, accounting for 37 per cent of British employment in this sector.

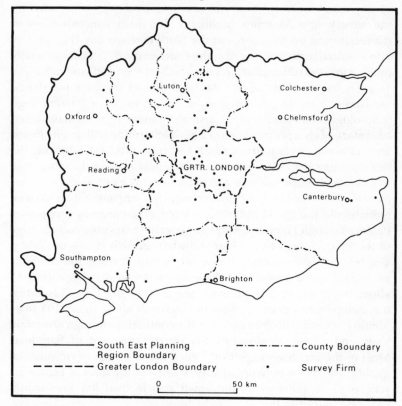

Fig. 5.2 The South East of England

5.2.3 *The San Francisco Bay area of California*

Evidence from the San Francisco Bay area economy reflects the more recent development of the region in general and of its industrial base in particular. Unlike Scotland, and to a lesser extent the South East of England, the Bay area has not been significantly hindered by the relics of once-prosperous Victorian industries now in decline. However, much of the region outlined in Figure 5.3 remains substantially undeveloped, which in its own way is a barrier to industrial growth.

As in the previous regions, the randomly sampled distribution of survey plants shown in Figure 5.3 strongly reflects the distribution of the total Bay area manufacturing industry. Apart from a small number of 'outlier' firms at, for example, Healdsberg in the north, Mariposa in the Sierra foothills to the east, and Monterey to the south, the bulk of the survey firm locations highlights the main concentrations of manufacturing on the shores of the San Francisco Bay (Figure 5.3). The industrial activity on or near the shore of the Bay can be generally divided into two distinct areas. First, the Berkeley–Oakland industrial area on the east side of the Bay is a district of older port-based industries, now largely in decline. In terms of a current high technology small firm sample, a small number of firms have taken advantage of cheaper premises of a poor environmental quality in this area. However, a car journey south-east from Oakland, following the Bay shoreline around to its southern extremity and then heading west, will create an impression of a gradual improvement in the environmental quality of the built environment accompanied by a general reduction in the age of industrial activity as the journey progresses. Passing through Hayward, Milpitas generally marks the eastern edge of the modern high technology industrial growth in Silicon Valley. The Silicon Valley area itself is a very compact industrial complex, measuring approximately 20 kilometres north-west–south-east by 10 kilometres north-east–south-west (see inset Figure 5.3). This industrial complex has developed south-eastwards along Route 101 from Menlo Park and Palo Alto in the north, continuing through Mountain View and Sunnyvale towards the older settlement of San José. Most of the buildings in Silicon Valley are of a high environmental quality. The dense network of survey firms in the inset of Figure 5.3 reflects the importance of this small area to total Bay area manufacturing output in the study industries.

From a statistical viewpoint, one benefit of the absence of a meso-level administrative area on which to base statistics is that the county-level data, compiled to provide a statistical basis for the region shown in Figure 5.3, offer a detailed sub-regional picture of the overall distribution of the two study industries. The most striking impression gained from Table 5.2 is of the dramatic changes that have taken place in the electrical and electronic equipment and instruments and related products sectors (SICs 3600 and 3800 respectively) since 1968. In particular, the growth in Santa Clara county, which includes Silicon Valley (inset, Figure 5.3), is remarkable. While employment in the county in the electronics sector SIC 3600 increased by 27,483 jobs in the decade 1968–78, growth in the instruments sector created a substantial 17,569 jobs from a base of 1,548 employees in 1968. The overall percentage growth for the whole Bay area in these sectors during the decade was 50 per cent for SIC 3600 and 791 per cent for SIC 3800. These figures were far in excess of the performance of

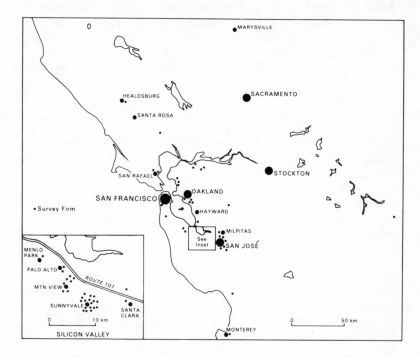

Fig. 5.3 The San Francisco Bay area of California

California as a whole (Table 5.2). It is clear from the inset in Figure 5.3 and the detailed county-level employment figures in Table 5.2 that Silicon Valley, within Santa Clara county, has provided the bulk of the rapid employment growth that has taken place in the decade 1968–78. This sub-regional expansion is an early aggregate indicator of the intense vitality and growth of this locality within the Bay area. Another feature of this decade of employment change in the Bay area is the mixed performance of San Francisco itself, with a picture of decline in SIC 3600 and a slight increase in SIC 3800 employment. Most other counties record modest increases in employment over the 10-year period.

5.3 The quality of the local physical environment

The following abbreviated plant-based survey data from firms in the study regions give evidence on the quality of physical conditions, notably factory conditions, and on the influence of the varying quality of such practical considerations on innovation in small high technology

Table 5.2 Employment in electrical and electronic equipment (SIC 3600) and instruments and related products (SIC 3800) for the San Francisco Bay area with disaggregated county-level statistics and totals, 1968–1978.

| | SIC 3600 | | | | SIC 3800 | | | |
| | | 1978 | | | | 1978 | | |
County	1968 Number	Number	Job change	% change	1968 Number	Number	Job change	% change
Alameda	2,920	3,376	+456		569	1,529	+960	
Contra Costa	1,524	812	−712		659	1,893	+1,234	
Marin	*	1,204	*		*	193	*	
Monterey	102	375	+273		—	147	+147	
Napa	—	60	+60		*	60	*	
Nevada	—	375	+375		—	—	—	
Sacramento	100	651	+551		—	352	+352	
San Benito	—	175	+175		—	—	—	
San Francisco	2,144	1,636	−508		173	416	+243	
San Joaquin	323	408	+85		—	—	—	
San Mateo	12,627	8,858	−3,769		277	1,108	+831	
Santa Clara	**34,739**	**62,222**	**+27,483**		**1,548**	**19,117**	**+17,569**	
Santa Cruz	*	1,248	*		—	175	+175	
Sonoma	197	653	+456		*	3,750	*	
Stanislaus	*	175	*		—	—	—	
Bay area total	54,676	82,228	+27,552†	+50†	3,226	28,740	+25,514†	+791†
California total	213,259	247,805	+34,546	+16	35,196	75,463	+40,267	+114

* Figure unavailable due to disclosure legislation.
† These figures are slightly exaggerated by the unavailability of very small employment figures for three counties in both sectors for 1968.
Source: US Bureau of Census (1968, 1978).

firms. Since the following tabulations are the first of many survey-based data in this book, it is an assistance to the interpretation of the strength of relationships contained within them to note that, where statistical testing is appropriate, a chi-square and probability value accompanies such data. The reader will be able to observe the exact level of significance from each table. The normal convention is that a probability of $p = 0.05$ or less implies a statistically significant difference between variables in the body of the table.

The local built environment of a region may influence innovation in two important ways. First, on a direct practical level, the quality and quantity of premises in the local area will either inhibit or enhance the ability of rapidly expanding innovative firms to obtain additional floorspace. Second, on a more subtle level, the visual appeal of a firm's factory building, and the aggregate visual impact of the wider industrial complex of which it is a part, may influence the firm's image. It is frequently the case that in the high technology firms of this study the promotion of high technology products is aided by a (pyscho-logically important) modern, bright factory which helps create the right impression with visiting customers and potential financial backers.

An obvious first step in the observation of the quality of a local regional environment is to consider the age of buildings occupied by respondent firms. Table 5.3, which indicates the age of factory buildings in the three study regions, displays some paradoxical, but not surprising, results. It is clear from the post-1970 category of newer buildings that Scotland and the San Francisco Bay area have the largest population of newer factories (i.e. 65 per cent and 49 per cent respectively). However, there are dissimilar causes for these similar regional effects. While the Bay area growth largely reflects the response of wholly private real-estate firms to a demand for high-quality new premises in the burgeoning Silicon Valley area, the Scottish high-quality new factories mainly stem from government assistance and direct-advance factory building. However, in many ways, the reason for the existence of a good stock of attractive new factories in any given local area is far less important than their existence *per se*.

The low proportion of post-1970 factories in the South East of England is largely explained by the various planning controls that have generally applied to this region throughout most of the post-war period. These controls have been particularly restrictive in Greater London (Oakey and Seaman 1981). The constraints of planning

Table 5.3 Age of factory buildings by region

Age of factory buildings	Scotland		South-East England		San Francisco Bay area	
(N = 173)	N	(%)	N	(%)	N	(%)
Pre-1950	15	(27.8)	28	(46.7)	6	(10.2)
1950–1959	2	(3.7)	6	(10.0)	4	(6.8)
1960–1969	2	(3.7)	14	(23.3)	20	(33.9)
Post-1970	35	(64.8)	12	(20.0)	29	(49.2)
Total	54	(100.0)	60	(100.0)	59	(100.0)

Chi-square = 41.81 $p = 0.0001$

legislation are implied by further evidence which indicates that 43 per cent of the survey firms in the South East stated that there was no room for expansion in their current plants as compared with a response of 29 per cent for Scotland and 30 per cent for the San Francisco Bay area. Moreover, the proportion of respondents that thought their present factory conditions poor was also higher in the South East, where a 28 per cent level in this category can be compared with 16 per cent and 5 per cent for Scotland and the Bay area respectively. The overall impression created by these data suggests that the availability of modern factory premises in the Bay area and in Scotland might be a marginal boost to innovation and growth in the survey firms in these regions. Conversely, it is clear that firms in the South East do experience problems with the available stock of premises. However, in terms of the technical progress of these firms, such a problem is more likely to retard innovation than be an absolute inhibitor.

5.4 Measures of regional innovativeness

This final contextual section immediately precedes a series of chapters that assume the hypothesised superior innovative performance of the South East of England and the San Francisco Bay area of California over Scotland. This section includes specific information on relative regional innovation levels not directly covered in subsequent chapters. It is important at this juncture to establish not only the direction of any bias in the three regional innovation levels, but also the measure of its extent.

The most significant and direct measure of regional technological progress in manufacturing firms has throughout previous research proved to be the incidence of product innovation (Oakey *et al.* 1982). It is therefore not surprising to discover that Table 5.4, which measures the regional incidence of new product innovation in the 5-year period prior to the survey, shows significant regional contrasts. In particular, the level of product innovation closely follows the 'expected' regional innovation performances of the study regions, with the San Francisco Bay area recording a very high 85 per cent level of innovation which is over 22 per cent higher than the Scottish development region's performance, with the South East of England in an intermediate position. However, this presence or absence measure, while drawing a valid sharp distinction between innovator and non-innovator, implies that all product innovations are of equal technological status, which is clearly not the case. Table 5.5. introduces a variable that ranks the

Table 5.4 The incidence of new product innovation by region

New product innovation (N = 174)	Scotland		South-East England		San Francisco Bay area	
	N	(%)	N	(%)	N	(%)
Innovation	34	(63.0)	47	(78.3)	51	(85.0)
None	20	(37.0)	13	(21.7)	9	(15.0)
Total	54	(100.0)	60	(100.0)	60	(100.0)

Chi-square = 7.84 $p = 0.019$

product innovations of Table 5.4 in terms of the extent of the technological change inherent in given products by region. Innovations were graded from the least-radical technological change of 'existing model improvement', through a category that identifies the product as a 'totally new product in a *similar* field of technology' to the firm's main previous production, to the most radical category, 'a totally new product in a *new* field of technology'. Two striking features emerge from this table. First, it is clear that, although the South East displays a high level of product innovation in Table 5.4, Table 5.5 indicates that 54 per cent of these innovations were in the least-radical technological shift category, compared with 32 per cent and 33 per cent for Scotland and the Bay area respectively. This less radical level

Table 5.5 Extent of technological change inherent in new product innovation by region

New product innovation (N = 121)	Scotland		South-East England		San Francisco Bay area	
	N	(%)	N	(%)	N	(%)
New model in existing product range	10	(32.3)	21	(53.8)	17	(33.3)
New product in a similar field of technology	4	(12.9)	10	(25.6)	17	(33.3)
Totally new product in new field of technology	17	(54.8)	8	(20.5)	17	(33.3)
Total	31	(100.0)	39	(100.0)	51	(100.0)

Chi-square = 11.97 $p = 0.018$

of technical 'shift' in the South East will be supported by subsequent data. Second, another feature that will reoccur is the notable fact that, although Scotland recorded the lowest level of product innovation in Table 5.4, it is also apparent from Table 5.5 that a high 58 per cent level of those firms that did innovate fell within the sub-sample of firms in the 'totally new product in a new field of technology' category. This performance is particularly impressive when compared with that of the South East of England (Table 5.5). Such results imply a heterogeneous sample of both non-innovators and technologically radical innovators for Scotland. This pattern of Scottish results is repeated below and in subsequent chapters.

Significantly, the regional incidence of process innovation in survey firms was much different. As in previous research, the coincidence of high levels of product innovation with prosperous regions was not repeated when the measurement focused on new process techniques (Oakey *et al.* 1982). Table 5.6 differs markedly from Table 5.4 in that the high level of innovation evident in the Bay area for product innovation is not repeated in the case of process innovation. Indeed, the Bay area records the lowest regional level of process innovation. Perhaps surprisingly, the South East records the highest level of process innovation, with a 69 per cent level of innovation compared with 50 per cent for Scotland and 48 per cent for the Bay area. The greater use of new process techniques in the South East of England might be a reflection of a tendency in the small firms in this region to operate in more standardised forms of production. This thesis gains support from the previously observed high proportion of South-Eastern firms claiming that their new product developments were merely new models of an existing product since, in comparison

Table 5.6 The incidence of new process innovation by region

New process innovation	Scotland		South-East England		San Francisco Bay area	
(N = 173)	N	(%)	N	(%)	N	(%)
Innovation	27	(50.0)	41	(69.5)	29	(48.3)
None	27	(50.0)	18	(30.5)	31	(51.7)
Total	54	(100.0)	59	(100.0)	60	(100.0)

Chi-square = 6.58 $p = 0.037$

with the problems of adapting process machinery to radical technical shifts, such updated models might be better suited to the introduction of standardised process machinery. These results will receive further support from Chapter 7 on research and development, where it is clear that a full-time commitment to research and development in the South East of England is lower than in other regions. Hence all these separate data suggest that although superficially the South East performs well in product innovation, the performance of this region at a more detailed level, beyond a mere presence or absence measure, is poorer than might have been expected. Conversely, the detailed innovation performance of Scotland proved to be more encouraging than initial results on the incidence of product innovation suggested.

6 Linkages and innovation

The principle of industrial linkage is fundamental to an understanding of the manufacturing process in individual manufacturing units. Chapter 4 has discussed the importance of linkages in creating agglomeration economies which produce broad impacts on the economics of industrial location. This chapter seeks to test many of the arguments of Chapter 4 and consider, with the aid of empirical evidence from the diverse study regions, the extent to which the quality of local linkages in the form of suppliers and customers influences the level of innovation in small independent high technology firms. However, it is initially helpful to explore the basic mechanisms of linkage at a conceptual level in individual small manufacturing firms.

6.1 Linkages and the small manufacturing firm

In terms of industrial linkage, the manufacturing firm is the hub of two separate sets of material flows from suppliers and to customers. These flows are maintained at varying distances from the factory, ranging from perhaps a few hundred metres in the case of a local firm in a nearby street, to thousands of kilometres when a foreign firm is involved. The distances of the linkage relationships, their profusion, and differences in the relative spatial diversity of inputs and outputs will vary between different forms of production and largely depend on the demands of the individual production process in which value is added to input materials. For example, while the input demands of ice manufacture may be satisfied by the nearest water main (Pred 1965), the assembly of complex electronic equipment might require inputs from many national and international locations (Oakey 1981). The relative ubiquity of essential inputs will be a strong determinant of input linkage distances in any given firm, while the location of its actual and potential customers will determine the length of output linkages. For while sufficient consumers of ice might be found within a short distance of the production plant, sophisticated products for high-performance aircraft might have specialised widely spread national and international customers. In general terms, it is likely that firms producing low technology products will develop relatively local input and output linkages, while technologically sophisticated products are more likely to display spatially diverse input and output linkage patterns.

The effect of this varying product sophistication on the distances of input and output linkages is best seen through an examination of diverse examples. For example, first consider a relatively low technology small firm manufacturing plastic buckets. From a technical standpoint, plastic buckets are a highly ubiquitous product on world markets. Hence, due to the large number of producers, profit margins need to be 'cut to the bone'. Thus there is little scope for expanding sales through a radical reduction in price. Also, there is little potential for specification improvement to stimulate sales. A relatively simple input and output linkage pattern emerges to suit this production regime. The major ingredients in the manufacture of plastic buckets are plastic granules. These may be purchased from a variety of producers but, due to the ubiquity of suppliers, the relatively low value of the raw materials, and the marginally significant impact of transport costs on tight profit margins, the bucket manufacturer will probably opt for a relatively local supplier. Apart from other sundry items, the plastic granules will comprise the bulk (by both volume and value) of the plastic-bucket manufacturer's material inputs. Since profit margins are low, it is often most economical to manufacture to order for a relatively local large wholesaler of a department-store chain. Thus, for the plastic-bucket maker, a simple overall linkage pattern might exist consisting of a relatively local main supplier and predominant customer. However, it is also probable that the customer would be relatively local because the large number of competing national and international producers would limit the development of exotic markets. None the less, the economic viability of production is facilitated by relatively low-cost local input and output flows which can be sustained at a low level of profitability.

If the other technological extreme of industrial production is considered, the mainly high technology firms of this study display markedly different linkage relationships from those of the plastic-bucket manufacturer. Most of the firms studied here neither purchase 'raw materials' in the true sense nor produce a finished product recognisable to the final consumer. Such high technology firms occupy production niches that are predominantly links in a chain of manufacture that stretches from true raw materials to the finished product, linking many electronic and instrument firms that work together to produce sophisticated finished products such as control systems for nuclear power stations or a jet fighter for international airforces. Most of the products of these small firms are of high value, since although the cost of a system of sensors for measuring the heat on the nose of a spaceship re-entering the earth's atmosphere might represent a fraction of the vehicle's total cost, this cost might nevertheless be measured in tens or hundreds of thousands of dollars. It is certain that such a sensory system will include thousands of individual components which may themselves be of a high technical specification and of high financial value.

The large number of input components to most high technology products requires many suppliers. The frequently high specification of such inputs also means that, in many cases, they may not be locally obtainable. This phenomenon has been noted in the British instruments industry (Oakey 1981). As compared with plastic-bucket making, the large number of inputs, their high specification and frequently exotic origin suggests that input costs will be relatively high. Taking the manufacture of the high-performance heat sensor as a comparison with bucket making, there is little scope for the sensor maker, as was the case for the bucket maker, to increase product sales through cost reduction. However, the reason for the sensor maker's inability to reduce costs is different from that of the bucket maker. While a degree of cost reduction through mechanisation is possible in areas of high technology industry where mass production is viable, such as calculator and electronic watch manufacture, high technology *small* firms do not, by definition, operate in such large production niches. In these small-scale areas of production, there is little scope for the substitution of cost-reducing machinery for skilled labour (Segal 1962; Oakey 1979b).

Economic viability for such producers is assured, however, through the specialist knowledge of conception and production inherent in the product. The relatively high transport cost of extensive and diverse supply and customer linkage relationships is met by the large amount of value added in production. But significantly, the value added in production is not so much the cost of input materials plus shopfloor labour and other sundry contributions to the manufacturing process, but predominantly the value added by the rather abstract input of human technical ability during development. It is true that any form of production will require a degree of expertise in the design of products. However, there can be little doubt that it is the technical input of human design expertise that contributes the bulk of value added to high technology production in small firms, thus paying for the expensive and spatially diverse input and output linkage relationships which ensure an adequate profit margin. Without this potential for high product prices facilitated by a superior or unique specification, the unit costs of production resulting from exotic linkage flows would in many cases be prohibitive. Unlike the plastic-bucket maker, there is virtually no price at which the high-performance heat sensor manufacturer may be forced out of the market, since there may be little or no alternative to his product. Indeed, if the national government is the sponsor of a product used for defence purposes, research and development costs may be paid in advance of production.

The two examples of production cited here were deliberately diverse, and the bulk of small firm manufacturing lies on a continuum between the two extremes. None the less, it is the contention of this chapter that the linkage flows of the high technology firms discussed in the following section will be more likely to resemble those of the heat sensor manufacturer than those of the maker of plastic buckets.

6.2 High technology firms, local linkage and innovation

The preceding arguments suggest, and empirical research into high technology industry confirms (Oakey 1981), that input and output linkages in high technology forms of production are generally large in number and national and international in distribution. Because of the relatively high value of inputs and outputs, transport costs are a small proportion of the total price to customers. Thus because of both insignificant transport costs and the frequently wide spread of suppliers and customers, it might be argued that, from a purely economic viewpoint, the availability of local suppliers and customers in an agglomeration is unlikely to enhance the innovation performance of high technology firms in that area, when compared with those in other locations.

However, an examination of Figure 6.1 proposes that the central small firm B is linked to its suppliers and customers not only by the overt flows of materials to and from the plant, but also by significant information 'feedback loops' to suppliers and from customers. These links may be particularly important in high technology small firms. On the input side, raw materials in Figure 6.1 are shown in inverted commas to imply that such materials are rarely raw materials in

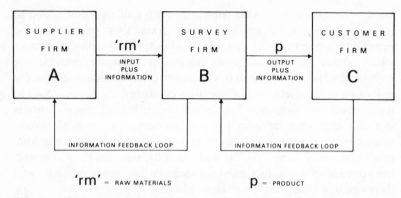

'rm' = RAW MATERIALS   P = PRODUCT

Fig. 6.1 Hypothetical material and information linkage flows

the sense of Weber's iron-production example (see Section 4.1). Indeed, most 'raw materials' in high technology industries are semi-finished components. Thus the supplier firm A might easily have been cited as the subject small firm B since most small firms, whether they subcontract or produce semi-finished products for general sale, are links in many chains of production which converge to produce large-finished products.

Although not confined to a single industry, this form of vertical disintegration of high technology production may link instrument, electronics and motor-vehicle manufacturers in a similar manner to the industry-confined production relationships noted by Wise (1949) for the gun and jewellery industries of Birmingham, Hall (1962) for furniture manufacture in London's East End and Martin (1966) for instruments in central London. This parallel is worthy of further pursuit because, as in the case of these now largely extinct intra-urban vertically disintegrated industries, close contacts throughs the information feedback loops are of paramount locational importance due to the 'behavioural' inertia they create, thus offering *real* opportunities for cost saving and technical innovation.

It should be restated that any given high technology firm will maintain a multiplicity of the links indicated in Figure 6.1, with a subsequent proliferation of information feedback loops to suppliers and from customers. Although materials which pass over long distances in such relationships may do so with little economic disadvantage, the effect of distance on innovation-related information feedback associated with material flows in a high technology chain of production might be more locationally important (Robertson 1974).

Other work has shown that great advantage lies in the close proximity of vendors and customers of services where close face-to-face contact promotes more frequent and fruitful information exchange (Goddard 1978). Hence, a worthy hypothesis might be advanced to argue that, while the direct economic advantages from the close proximity and interchange of materials between high technology firms are insignificant, the longer-term more subtle advantages gained from information exchange through local 'feedback loops' significantly assist innovation. This line of argument leads back to the main hypothesis of this book that variations in the quality of the local industrial environment may influence small firm innovation levels. Clearly, the following analysis of three diverse industrial regions might expose such differences if they occur.

Although it is valid to talk of linkages in terms of chains of production, the questionnaire interview method of data collection clearly sets the interviewed firm in the position of subject B in Figure 6.1. Hence, since this study looks at the behaviour of firm B and not at a complete linkage chain, input and output linkages are clearly defined and have a precise subject on which to exert their separate effects. Thus input and output linkages and their spatial effects on survey firms will be dealt with in turn.

6.3 Input linkages and innovation

Following the preceding argument on the possible importance of information flows in material linkage interaction, it might be argued that, if local input linkages generate information which aids innovation, there would be greater local sourcing of inputs in the highly innovative survey region of the San Franciso Bay area than in Scotland. For while it might be argued that links with customers are more important to the small firm innovation process than links with suppliers, due to the feedback of information directly related to products, a sophisticated and locally available set of subcontractors and stockists might none the less clearly assist innovation. This may be particularly true in the context of this study, since the continuing development of new products in the high technology small firms examined in this section creates the need for a constant stream of new or improved components and materials, due to the relatively short product life cycles. In such circumstances, physically close liaison with specialist suppliers may be an invaluable spur to innovation. The extent to which survey firms sourced their material purchases from local suppliers (within a 30-mile radius) might establish a link between regional differences in innovation and the extent of local input sourcing. It is worth reiterating that the evidence of Chapter 5 indicates that firms in the San Francisco Bay area were significantly more innovative than their Scottish counterparts.

It is clear from Table 6.1 that there is a significant difference between the local-purchasing expenditure of Bay area firms and that of their Scottish counterparts. While a mere 11 per cent of Scottish firms purchased over 50 per cent of their material inputs (by value) locally, the equivalent figure for Bay area firms was 68 per cent, while the survey firms in the South East of England recorded an 18 per cent level. This sharp regional contrast is confirmed by a chi-square test on Table 6.1 that easily exceeds the $p = 0.001$ level of significance. Clearly, the regions differ structurally in several respects that may partly explain this pattern of results. Ironically, the low level of local sourcing in the South East of England is in part caused by the relative size of this region. For although the South East is minute in comparison with the United States or even California, it is large when compared with the concentration of industry in the San Francisco Bay area. Hence, many South-Eastern suppliers would not have fallen within the 30-mile radius encircling the survey plant, and have thus been excluded from Table 6.1. However, the very high incidence of local sourcing in the Bay area is valid evidence of a physically intense concentration of essential inputs. This high level of local sourcing bears witness to

Table 6.1 Percentage purchased locally (within a 30-mile radius) by region

Percentage purchased locally (N = 174)	Scotland		South-East England		San Francisco Bay area	
	N	(%)	N	(%)	N	(%)
Nil	8	(14.8)	10	(16.7)	7	(11.7)
1–24	34	(63.0)	27	(45.0)	8	(13.3)
25–49	6	(11.1)	12	(20.0)	4	(6.7)
50–74	2	(3.7)	7	(11.2)	17	(28.3)
75–99	3	(5.6)	4	(6.7)	22	(36.7)
100	1	(1.9)	0	(0.0)	2	(3.3)
Total	54	(100.0)	60	(100.0)	60	(100.0)

Chi-square = 59.35 $p = 0.0001$

the range and quality of high technology suppliers in this tight-knit industrial concentration. Indeed, it should be remembered that the Silicon Valley firms, which comprise 60 per cent of all Bay area firms, are contained within a rectangular area that measures no more than 20 by 10 kilometres. Moreover, beyond the obvious physical convenience of such sourcing arrangements, the subtle technical information benefits for innovation from such close liaison with suppliers are likely to contribute to the high level of Bay area product innovation observed in Chapter 5.

There might be a link between the level of regional product innovation and the propensity of high technology small firms to source inputs locally. However, this possible association may be tested in a different manner by examining the extent to which survey respondents found that their local supply relationships inhibited innovation. To this end, survey firm executives were asked if a poor level of local services to their production process, defined in terms of material supplies, subcontracting, and services to production machinery, directly inhibited innovation in their firm. Table 6.2 indicates that, of the 19 firms acknowledging this problem, 14 were located in Scotland, comprising 26 per cent of the Scottish total compared with 2 per cent for the South East and 7 per cent for the Bay area.

Table 6.2 Extent to which poor local supply linkages inhibited innovation

Effect of local supply linkage on innovation	*Scotland*		*South-East England*		*San Francisco Bay area*	
(N = 174)	N	(%)	N	(%)	N	(%)
Inhibition	14	(25.9)	1	(1.7)	4	(6.7)
None	40	(74.1)	59	(98.3)	56	(93.3)
Total	54	(100.0)	60	(100.0)	60	(100.0)

Chi-square = 18.90 $p = 0.0001$

Thus, one in four of the Scottish firms claimed that poor local input linkages had inhibited its innovation performance. The cause of the problem was attributed almost equally to a poor choice of stockists and to deficiences in local subcontracting facilities. These results reinforce those of Table 6.1 in that the lower use of local material inputs by Scottish firms in Table 6.1 must partly be explained by the inhibitions noted in Table 6.2. These combined results imply that a good set of local suppliers may aid innovation in the small high technology firm, while a poor level of local input services may depress innovation.

However, the impact of local linkage quality on innovation in small high technology firms is not straightforward. For example, it would be an obvious exaggeration to claim that a rich local supply of material suppliers is the major cause of the high innovation levels in the San Francisco Bay area of California. None the less, there is clear circumstantial evidence in the above data of concomitance between high levels of regional product innovation and a tendency to source inputs locally and be satisfied with their quality. In an attempt to establish a direct link between product innovation and local input sourcing, Table 6.3 relates the incidence of product innovation with the extent of local purchasing. Interestingly, there is a marginally greater tendency for firms purchasing over 50 per cent of their inputs (by value) locally to produce a higher 83 per cent of product innovation than the 72 per cent level recorded by those firms with predominantly non-local sources of input linkages. The significance of a chi-square test on Table 6.3 equated $p = 0.133$, fractionally outside the normally accepted $p = 0.05$ level. This marginal level of significance perhaps reflects the differing effects of the variable quality of local supply linkage on firms with different approaches to innovation.

For example, it is notable that in Table 6.2, the 14 Scottish firms acknowledging the inhibiting effects of poor local linkages are divided

Table 6.3 Extent of local purchasing by incidence of product innovation

New product	Extent of local purchasing			
	< 50%		> 50%	
(N = 174)	N	(%)	N	(%)
Innovation	84	(72.0)	48	(83.0)
None	32	(28.0)	10	(17.0)
Total	116	(100.0)	58	(100.0)

Chi-square = 2.35 $p = 0.133$

almost equally into 8 product innovators and 6 non-product innovators. This result suggests that poor local input linkages may have two simultaneous effects, depending on the firm concerned. First, they may inhibit, but not arrest, the innovation process in aggressively innovative firms, reflecting the added stress such firms must put on the poor resources of their local area. Second, poor local input supplies may help to perpetuate a non-existent level of innovation in the non-innovative firms of the same area. Conversely, Silicon Valley, with its excellent range of local input materials, must marginally enhance innovation in innovative firms and perhaps help stimulate innovation in firms that, were they in Scotland, might have remained non-innovators. Thus, while the quality of local material input linkages is unlikely to be a main determinant of innovation, it must have a marginal influence on innovation as a contributory part of the total agglomeration effect of many combined local resource advantages discussed in later chapters.

6.4 Output linkages and innovation

The traditional view of output linkages is that they are less likely to exert locational influence on the economics of production. Weber, when considering heavy industries that reduce the weight of inputs during production, considered material orientation to be a dominant feature due to weight loss (Weber 1909). Even for more modern forms of production, both the value added in production and the potentially diverse distribution of customers must tend to dissipate the locational constraints on sales relationships (Lever 1974), with the high value of outputs ensuring the economic viability of transportation over long distances, particularly for high technology outputs (Oakey 1981).

However, Figure 6.1 proposes that for every outward flow of materials there is a backwash of information from the customer. The probable absence in local output linkages of any strong advantage based on transport costs does not negate the possibility of subtle local technical information bonds that might lock a small firm into a particular technical-information-rich environment. Again, these arguments benefit from an ability to resort to the survey data set. It might be expected that if agglomeration economies ensuing from output linkages offer either direct cost savings or more subtle technical information advantages that aid innovation, they will be detectable in the highly innovative firms of the San Francisco Bay area.

In a similar manner to the approach adopted for input linkages, respondents in the three study regions were asked to estimate the percentage of their sales (by value) despatched to customers within a 30-mile radius of their plants. Table 6.4 indicates that there is little evidence of a predominantly local orientation of customer links in any of the study regions. For example, the percentage of firms with over 50 per cent of their sales in the local area was for Scotland, the South East of England and the San Francisco Bay area, 21 per cent, 27 per cent and 33 per cent respectively. Notably, the difference in these figures between Scotland and the Bay Area is not great, and unlike the equivalent table for input linkages (Table 6.1), where the chi-square test exceeded the $p = 0.001$ level, the value for Table 6.4 was an insignificant $p = 0.31$.

Table 6.4 Percentage sold locally (within a 30-mile radius) by region

Percentage sold locally	Scotland		South-East England		San Francisco Bay area	
(N = 173)	N	(%)	N	(%)	N	(%)
Nil	10	(18.9)	4	(6.7)	8	(13.3)
1–24	28	(52.3)	30	(50.0)	27	(45.0)
25–49	4	(7.5)	10	(16.7)	5	(8.3)
50–74	4	(7.5)	5	(8.3)	10	(16.7)
75–99	7	(13.2)	11	(18.3)	9	(15.0)
100	0	(0.0)	0	(0.0)	1	(1.7)
Total	53	(100.0)	60	(100.0)	60	(100.0)

Chi-square = 11.50　　　　$p = 0.311$

Although these initial results tended to suggest a low local locational importance for sales linkages, respondents were questioned in greater detail on their customer relationships. Firm executives were asked if their company possessed a single major customer that purchased over 10 per cent of their total output by value. Those firms acknowledging such a customer represented 61 per cent, 62 per cent and 65 per cent of the survey firms in Scotland, the South East of England and the San Francisco Bay area respectively. Given the frequent protestations of many interviewees that, if at all possible, sales were distributed to as wide a range of customers as possible to avoid the risk of becoming dependent on a single customer, these regional percentages are high. However, only 18 per cent of all survey firms sold over 50 per cent of their output to a single customer. One problem for many producers is that although they would prefer to distribute their output as widely as possible, the recent deep recession in both Britain and America has forced many small firms to sell to any customer with a demand, regardless of his current level of purchase.

Table 6.5 Location of major customer purchasing over 10 per cent of a firm's output by region

Location	Scotland		South-East England		San Francisco Bay area	
(N = 104)	N	(%)	N	(%)	N	(%)
Local (within a 30-mile radius)	11	(32.4)	12	(32.4)	15	(45.5)
Within region or state	7	(20.6)	12	(32.4)	6	(18.2)
Within nation	12	(35.5)	13	(35.1)	9	(27.3)
Abroad	4	(11.8)	0	(0.0)	3	(9.1)
Total	34	(100.0)	37	(100.0)	33	(100.0)

Chi-square = 7.25 $p = 0.298$

None the less, Table 6.5 reinforces the evidence of the low general importance of local customers by indicating that the locations of major customers purchasing over 10 per cent of output from firms acknowledging such patrons were not particularly local (i.e. within a 30-mile radius). Again, the San Francisco Bay area shows a slightly higher tendency to maintain important local customers, with 15 firms (25 per cent of the Bay area total) acknowledging an important local customer. But, in common with Scotland and the South East of England, the majority of the output despatched to important customers is sent either to the rest of the region or to national or international destinations. This result tends to suggest that neither important customer relationships nor volume of materials are particularly inhibited by distance, thus detracting from any argument that local output linkage economies exist to encourage innovation in Silicon Valley. The information on exports in Table 6.6 supports the general trend of disparate sales linkages by indicating that a substantial majority of survey firms in all regions export their outputs. While the export performances of the South-Eastern and Bay area

Table 6.6 Percentage exported by region

Percentage exported	Scotland		South-East England		San Francisco Bay area	
(N = 174)	N	(%)	N	(%)	N	(%)
Nil	24	(44.4)	20	(33.3)	20	(33.3)
1–24	14	(25.9)	26	(43.3)	31	(51.7)
25–49	13	(24.1)	6	(10.0)	8	(13.3)
50–74	1	(1.9)	7	(11.7)	1	(1.7)
75–99	1	(1.9)	1	(1.7)	0	(0.0)
100	1	(1.9)	0	(0.0)	0	(0.0)
Total	54	(100.0)	60	(100.0)	60	(100.0)

Chi-square = 20.67 $p = 0.024$

firms are almost identical, Scotland, with the highest proportion of firms in both the 'no exports' category and the 25-49 per cent range, again records a heterogeneous performance. This mixed picture is similar to earlier and subsequent results on Scotland. The generally high level of exports must, at least in part, be related to the generally high technology nature of the small firms in this survey and to the transportability of high technology industrial output (Oakey 1981).

Another dimension to small firm output linkages is the sectors into which the output of survey firms flows. Do the study firms in the survey regions serve a broad spectrum of industries, or are their outputs narrowly directed? Firms were asked to state whether they sold over 25 per cent of their output to any particular industry, defined for convenience at the British industrial order-heading level. The response to the question was predominantly affirmative, with the San Francisco Bay area firms showing the greatest propensity with an 82 per cent level of sales to a single industry, while the South East of England and Scotland were close behind with incidences of 80 per cent and 78 per cent respectively. This high level of sectoral specialisation is examined further in Table 6.7, where it is clear that the range of industries supplied by these specialist firms is very small. Indeed, of the 139 firms acknowledging the despatch of over 25 per cent of their output to a single industry 87 (65 per cent) claimed that their output was dispatched to only three manufacturing sectors; these were: electronic engineering (48 per cent), defence (9 per cent) and instrument engineering sectors (8 per cent).

Table 6.7 Major customer industries based on British industrial order headings by region

Industry (N = 139)	Scotland		South-East England		San Francisco Bay area	
	N	(%)	N	(%)	N	(%)
Electrical engineering	20	(47.6)	20	(41.7)	27	(55.1)
Armaments & defence	4	(9.5)	5	(10.4)	4	(8.2)
Professional & scientific services	3	(7.1)	5	(10.4)	3	(6.1)
Instrument engineering	1	(2.4)	5	(10.4)	4	(8.2)
Mechanical engineering	1	(2.4)	6	(12.5)	1	(2.0)
Vehicles	3	(7.1)	2	(4.2)	2	(4.1)
Other	10	(23.8)	5	(10.4)	8	(16.4)
Total	42	(100.0)	48	(100.0)	49	(100.0)

However, the spatial distribution of the industries obtaining over 25 per cent of this sub-sample's output is similar to earlier results, indicating a low emphasis on local customer industries. Table 6.8 again shows that the San Francisco Bay area has the highest number of firms with local linkage with such industries, although these firms comprise only 31 per cent of the sub-sample and 25 per cent of all Bay area firms. A most striking feature of Table 6.8 is the low level of nationally based customer industries for Bay area firms, which reflects a high level of

Table 6.8 Location of industry purchasing over 25 per cent of output by value

Location (N = 139)	Scotland		South-East England		San Francisco Bay area	
	N	(%)	N	(%)	N	(%)
Local (within a 30- mile radius	9	(21.4)	8	(16.7)	15	((30.6)
Within region or state	7	(16.7)	7	(14.6)	5	(10.2)
Within nation	20	(47.6)	25	(52.1)	14	(28.6)
Abroad	6	(14.3)	8	(16.7)	15	(30.6)
Total	42	(100.0)	48	(100.0)	49	(100.0)

Chi-square = 9.89 $p = 0.129$

either local or international sales linkages. Indeed, if the local and international industrial links are combined for the three study regions, the proportion for the San Francisco Bay area equals 61 per cent compared with 36 per cent and 33 per cent for Scotland and the South East of England respectively. The sum effect of the results seems to suggest that there is considerable industrial sectoral concentration in most survey firms, but that this sectoral specialisation is spatially diverse and not located in the local area of survey firms.

6.5 Evidence of technical information feedback from customers

The following analysis seeks to investigate in empirical detail the hypothesised existence of a significant 'feedback loop', or backwash, of information from customers to survey firms. Clearly, the existence of the feedback of important information instrumental to new product design would be of particular advantage to survey firms, although the preceding evidence suggests that if such links exist, they are unlikely to be local in origin since there is no preponderance of local customers in survey firms. There are various levels of quality inherent in any contact between survey firms and customers, ranging from the occasional feedback of informal verbal advice on the performance of products, to the drawings and specifications of individual products or sub-assemblies common in typical subcontracting relationships.

Initially, survey firms were asked if customers provided any technical information on product design. A significant number of firms in all regions acknowledged the passage of information on product design, with Scotland and the South East of England recording a 63 per cent affirmative level and the San Francisco Bay area a slightly lower level of 55 per cent. In response to a further, more detailed question asking if there existed a particularly important customer that provided information on product design, Table 6.9 indicates that a lower 41 (24 per cent) of the total survey sample claimed an important technical feedback link with a single customer, although in the instance of Scotland this proportion was a higher 31 per cent of the Scottish total and 52 per cent of the subsample in Table 6.9. However, further investigation revealed that only 16 firms acknowledged the passing of product specifications; the remaining links of lesser importance included the passing of drawings, research and development collaboration, and verbal contact in person and over the phone. Only in Scotland and the Bay area was there any evidence

Table 6.9 The incidence of customers providing important technical information by region

Customers	Scotland		South-East England		San Francisco Bay area	
(N = 103)	N	(%)	N	(%)	N	(%)
Important customer	17	(51.5)	12	(31.6)	12	(37.5)
None	16	(48.5)	26	(68.4)	20	(62.5)
Total	33	(100.0)	38	(100.0)	32	(100.0)

Chi-square = 3.03 $p = 0.219$

that the customers stimulating these important technical information links were local. While at this stage sample numbers become few due to disaggregation, 4 of the South-Eastern survey firms in Table 6.9 claimed their technical information link to be local, compared with 8 for Scotland, while 8 of the 12 Bay area firms had local contacts. The overall results on customer information feedback, however, suggest that the direct passage of technical information from local customers to high technology small firms is generally not significant and unlikely to be a widespread aid to innovation. Such a result detracts from any information linkage agglomeration effect that might have been provided by local customers.

6.6 A new interpretation of linkages and local agglomeration economies

In keeping with the traditional convention when considering linkages, this chapter has divided its analysis of material flows into sections on input and output linkages. Consideration of linkages in this dichotomous manner stems from Alfred Weber's simplified model of industrial production in which the production plant was interposed between raw materials and the market (Weber 1909). In such simple circumstances, where the plant considered was the sole unit in the production process (narrowly defined as iron making), it was valid to term material flows as inputs and outputs. However, since Weber's early writings, industrial production has generally become far more complex. It is now appropriate to view the passage of raw materials to the market as a chain of production which may involve many individual linked plants within a company, or indeed a collection of independent firms. Such chains of production will be many in number, converging on the plant of final assembly.

Importantly, this 'production chain' approach to linkages, with the consequent introduction of many production plants within the chain, changes the nature of linkages since, when they are viewed in terms of the destination plant of final assembly, they are *all* input linkages to the point of final assembly. Viewed from this angle, the converging intermediate sub-assemblies and the plant of final assembly become focuses, and not individual plants involved as links in the chain. The main advantage of this revised means of viewing linkages is that it emphasises that there is no difference in principle between input and output linkages, since the input linkage of one firm's input is the output linkage of another firm in the chain of production.

This new approach permits an interpretation of the data in this book. It has been noted that the highly prosperous and innovative firms in the San Francisco Bay area source a very high proportion of their inputs locally, and that a significant minority of firms in the depressed Scottish development region indicated that input shortages in both subcontractor and component services inhibited their level of product innovation. Conversely, local sales from survey firms did not appear of particular locational significance in any of the regions studied. It appears, then, that local inputs to survey firms are important in the most prosperous and innovative region, while local sales are of low significance to innovation performance in all the regions. The linkage chain approach offers a useful explanation of this general pattern of results.

In answer to the question on important technical links with customers, many respondents throughout the study regions forcefully made the point that although the customer sets the problem, the study firm solves it with internal skills. Such an exercise frequently results in a new product for the survey firm, but importantly, while the customer may have stimulated the innovation through the suggestion of a new product niche, the expertise used to solve the problem was *internal* to the survey firm. This observation is an important clue to an understanding of the survey results and the relevance of the current linkage chain approach to material flows.

The simplified three-stage linkage chain shown in Figure 6.2 indicates that as materials flow forward along the chain towards the plant of final assembly, technical dependency flows in the reverse direction. If, as might well be the case, firm B in Figure 6.2 is taken as

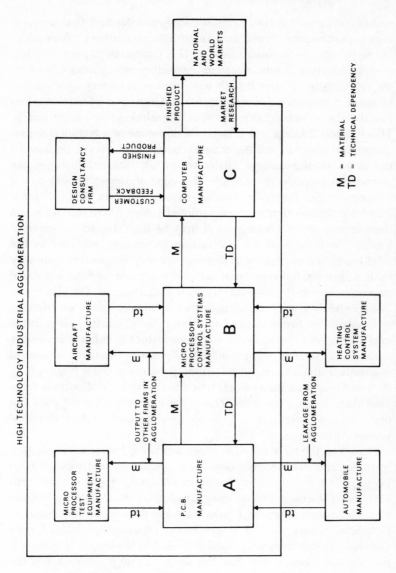

Fig. 6.2 Material flows and technical dependency in a high technology industrial agglomeration

one of the current survey firms, it both generates technical dependence in the computer firm C and is in turn the technological dependant
of the printed circuit board firm A. This pattern of dependency has
obvious effects of significance to the preceding survey results. Because
the relationship of firm B with its customer is one in which it has
technical dominance, the necessity for proximity and interaction is less
because the technical basis for the relationship is internal for firm B.
This is a relationship that firm B can dominate in a technical sense,
although it might be smaller than its customer. However, because of
the flows in the linkage chain, where all material linkages are
considered identical in principle, the same argument applies to the
instance where firm B is itself a customer of firm A. Even for a
relatively simple item such as a printed circuit board, the technical
dependency of firm B on firm A may be high due to the external
location in firm A of the production technology inherent in the
relationship. Hence, technical dependency may thus explain both why
local customer linkages from survey firms are of low locational
importance, and why the evidence on the significance of local supplier
linkages tends to suggest that they are locationally important, offering
advantages for firms situated in their locality. Clearly, since firms
export a significant proportion of their output to other parts of the
nation and abroad, there will be substantial 'leakage' of output
materials from all linked plants in the production chain (Figure 6.2),
thus ensuring that the agglomeration is an exporter of outputs from
each stage in the production chain. For example, Silicon Valley not
only exports computers, but also printed circuit boards and microprocessor control systems.

However, the technological dependencies of Figure 6.2 are highly
complex, both in time and space. The simplified chain depicted would
be, in reality, much longer. For every computer-manufacturing firm
C, there will be many linkage chains, with this firm as their destination.
Continued maintenance of links will also depend on the mutually
beneficial nature of the dependency relationship. Indeed, this
dependency is very tenuous and is limited to the technical sphere. If
the computer firm C sets firm B a problem that it cannot solve, the
material flow may lapse. It is also possible that firm B will decide to
manufacture printed circuit boards internally or purchase them from
elsewhere, thus negating its technological dependence on firm A. Set
in a broader context, the tenuous nature of dependency between high
technology firms is reflected in another manner. Although the
computer firm C in Figure 6.2 may be technically dependent on firm B

for a particular component, if the product of firm B achieves high sales, both to firm C and other consumers, firm C may decide to attempt the acquisition of firm B through the immediate or gradual purchase of equity (Channon 1973; Smith 1982). While this objective may or may not be achieved, the technological superiority enjoyed by high technology small firms over larger customers is not universal and does not extend to capital strength. Indeed, in some cases this small firm technological superiority can be dangerous if it provokes acquisition.

It is also important to stress that this phenomenon is most common in high technology industry. The complexity of production in high technology firms frequently means that specialist product niches exist which are best-filled by small high technology firms. Other simpler forms of production may not display the previously described technical dependencies. For example, an engineering firm sub-contracting the manufacture or machining of metal components to a smaller firm would technically dominate this supplier due to the technical ubiquity of the service and the number of potential suppliers.

The concept of linkage chains of production and technical dependence offers an explanation for the observed differences in the incidence of local input and output linkages and their effects on innovation in small high technology firms. Adequate material linkage support to production is obviously an important factor in determining the best location for high technology industrial production, and indeed, is always a priority in determining a suitable location for mobile high technology production on the rare occasions that this occurs (Oakey 1981). These results tend to re-emphasise this view in general and to suggest also that the San Francisco Bay area of California in particular offers advantages conducive to the growth of small high technology firms through its rich and spatially concentrated network of material suppliers at all stages of high technology production. Indeed, in a final 'open-ended' question to all survey respondents that enquired what was the greatest aid to innovation in the firm, 8 Bay area executives (14 per cent) mentioned the quality of local suppliers, while none of the British executives in Scotland or the South East of England mentioned this advantage. However, taking a broader view, this local input resource advantage must be viewed as a partial cause of any total local agglomeration stimulus to innovation in the Bay area, in which the total agglomeration advantage might be seen as greater than the sum of its constituent parts.

7 Research and development and innovation

Apart from those small firms engaged in subcontracting, where items of production are specified to them by their customers, small manufacturing firms are predominantly engaged in the production of an indigenously designed finished or semi-finished product of some description. Given the propensity, previously discussed in Chapter 4, for products in high technology small firms to enjoy relatively short life cycles, it is clear that sooner, rather than later, existing products must be improved or replaced to ensure continued sales and the medium-term growth of the firm. Moreover, it is also evident that in order for such periodic revisions in product specification to be made, a firm must possess internally, or have access to, the technical capacity to enact specification improvements. This technical capacity is normally a research and development facility, although it may exist at various levels of commitment and be drawn from various sources. Most small firm executives would not identify their research and development efforts as 'pure' research, but would rather describe them as development work since, in the main, they are problem oriented and incremental rather than directed at the pursuit of new technical knowledge *per se*. High technology small firm executives generally view their research and development effort as an expensive necessity which must be closely linked to measurable product improvement in order to justify the input of scarce resources. For such expenditure has no short-term return and uncertain medium-term benefits, thus differing from much production-oriented expenditure.

7.1 Investment cycles, research and development cycles and innovation

The problem of shortage of investment capital is common in all successful small firms. However, a pervasive cause of investment problems associated with growth in *high technology* small firms is the cyclical nature of both product sales and research and development expenditure. Most high technology small firms are founded on a main product. The growth and subsequent decline of sales from an initial or subsequent new product imply that the profits will not be uniform. This cyclical revenue will thus detract from the overall security of the firm and from its creditability to those providing risk capital. Moreover, to exacerbate matters, it is likely that maximum research and development expenditure takes place in the periods prior to a new product launch when profits are at a low ebb. This phenomenon is shown diagrammatically in Figure 7.1 where a small high technology firm with an aggressive approach to research and development investment is hypothesised, implying an expensive full-time research and development commitment. Profits are defined as sales and service revenue minus production costs which, for the purposes of the current argument, *do not* include research and development expenditure, which is indicated separately.

In this conceptualisation, the birth of the firm coincides not only with low or non-existent profits, but also with high research and development costs on an emerging product. Moreover, this initial crisis repeats itself at intervals during the life of a firm, depending on the individual product life cycle concerned. This simplified model assumes that all other performance-influencing variables, such as the performance of the economy, are held constant. Points of financial stress occur in Figure 7.1 when research and development expenditure exceeds profits. For a short period it is necessary to obtain an overdraft or other loans with which to 'weather the storm'. In this sense, borrowing may be seen as a rather expensive means of smoothing the profit curve of the firm, since future sales of a new product will, hopefully, finance the debts incurred during the stress periods caused by intense research and development. This problem of cyclical revenue is further compounded by the impact of taxation on the ensuing 'lumpy' profits. While product life cycles will vary between firms, in this conceptual generalisation a product life cycle of approximately five years is adopted, a cycle duration common in high technology small firms. For at least two of these five years, in mid-life cycle, profits are high, thus attracting a high rate of taxation. However, during a period of loss or low profits there is no return of the tax that would have been deemed overpaid if averaged over the whole five-year period. The net result of the combined negative factors of high interest on loans and taxation on profits is that the small firm has less money to spend on research and development directed at the development of future products.

Perhaps the biggest dilemma facing small-firm owners relates to risk. Seen in a retrospective light, the success of the first and second

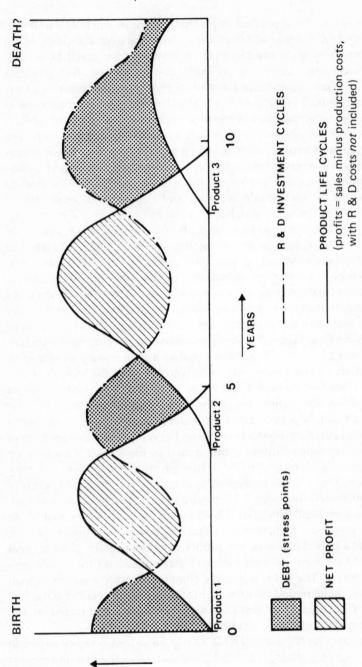

Fig. 7.1 Research and development investment and product life cycles in the small high technology firm

products in Figure 7.1 almost implies some form of symbiotic relationship between research and development effort and subsequent sales. However, lacking a crystal ball, the individual owner has no means of knowing the level of future sales when considering the appropriate level of research and development investment for any given new product development. However, he does know the level of past expenditure on a product development at any given moment in the investment cycle. Since most small firms cannot afford the luxury of market research, much of the decision making on how much to invest in a project and when to stop will stem from past experience and a knowledge of current and future market conditions. While, by definition, all surviving firms manage to ensure that research and development effort is broadly commensurate with subsequent sales potential, it is always possible that, as in the case of product life cycle (3) in Figure 7.1, subsequent sales do not offset research and development expenditure, thus in all probability causing the demise of the whole firm. This phenomenon emphasises that there is no justification for the argument that research and development investment is desirable at any price. This concept will be further developed in Chapter 9, which considers investment capital.

Hence, a wide range of approaches and subsequent commitments to research and development is common in small firms. Clearly, both firm size and the sophistication of the product technology will influence the level of research and development commitment. This level must also reflect, at least in part, the overall level of ambition inherent in the management approach (Senker 1979). As a general rule of thumb, however, the quality of the technical information derived from research and development initiated in small firms is proportional to the total level of capital investment. For example, the most rewarding technical information on improved product design is likely to ensue from a relatively expensive internal research and development facility containing full-time researchers. Higher cost notwithstanding, it is clear that the fruits of such efforts are the confidential property of the firm concerned, a factor which may be important in competitive high technology markets (Oakey 1981). At a much lower level of commitment, a small firm might employ an external consultant on a contract-by-contract basis. This approach is clearly more flexible in that this source of expertise does not involve a full-time research and development burden on employment resources. In extreme cases, a small firm might rely solely on such a consultant, but more frequently, will use this service to support any given level of internal effort.

The local availability of such innovation support services would be a clear advantage to innovative firms and is thus an issue of central concern to this book. However, although flexible, problems of confidentiality may arise with respect to external consultants and their work. Moreover, the technical output from such a relationship will not generally yield the same standard of results as that derived from an internal research and development department (Oakey 1979c). Since the varying levels of research and development commitment imply widely different approaches to innovation in survey firms, the various approaches are considered separately in detail in the following section.

7.2 Internal research and development and innovation

Internal research and development facilitates the interactive linking, under one roof, of development effort to the specific production needs of the firm. In many small high technology firms the chief executive or owner is a qualified engineer. Hence, it is likely that the *raison d'être* of the firm was the initial product idea of such an individual. Thus, for good or ill, there is rarely much conflict between specific technical research and development objectives and the wider aims of the firm, since the key decision maker is frequently the technical innovator (Senker 1979, 1981). Virtually all high technology small firms will make an effort to preserve their competitive technological edge through some form of evolutionary research and development. This assertion is borne out in Table 7.1 which conclusively indicates that only 26 (15 per cent) of the 174 survey firms did not maintain some form of internal research and development facility. Moreover, the

Table 7.1 The incidence of internal research and development by region

R & D effort	Scotland		South-East England		San Francisco Bay area	
(N = 174)	N	(%)	N	(%)	N	(%)
Internal R & D	42	(77.8)	51	(85.0)	55	(91.7)
No R & D	12	(22.2)	9	(15.0)	5	(8.3)
Total	54	(100.0)	60	(100.0)	60	(100.0)
Chi-square = 4.31		$p = 0.116$				

asserted link between internal research facilities and innovativeness, evident in earlier work (Thwaites *et al.* 1981), is underlined by Table 7.2, which shows that of the subsample of survey firms recording a product innovation during the five-year period prior to survey, 93 per cent had internal research and development facilities, while only a very small minority (7 per cent) had no such facilities. From a regional viewpoint, it is significant that the largest number of firms with no research and development effort were located in Scotland.

Table 7.2 Presence of internal research and development by incidence of product innovation

R & D effort	Product innovation		No innovation	
(N = 174)	N	(%)	N	(%)
Internal R & D	123	(93.2)	25	(59.5)
No R & D	9	(6.8)	17	(40.5)
Total	132	(100.0)	42	(100.0)

Chi-square = 28.40 $p = 0.0001$

However, the mere existence of research and development on site is an imprecise category which aggregates various degrees of full- and part-time commitments. Thus it is useful at this point in the discussion to consider full- and part-time research and development separately, since these approaches imply different organisational structures and varying levels of financial commitment within individual firms.

7.2.1 *Part-time research and development*

Part-time research and development is particularly common in small firms. It is a compromise between the need for innovation through research and development and the high cost of full-time research and development staff. Since in many small firms the owner is frequently the production engineer, accountant and salesman, great conformity of purpose is achieved by the owner's performance of development work in his spare time (Freeman 1974). Indeed, in certain instances, this practice precedes the formal establishment of the small firm. Work on the firm's initial product frequently takes place during a period of previous employment. Since the employment of full-time research and development workers is often a prohibitive expense, the part-time efforts of the owner, plus perhaps the technical director and production engineer, may be an optimal level of effort from a financial viewpoint. However, while part-time research and development would record a lower research and development cost cycle in Figure 7.1, the resultant output as reflected in new product sales would be similarly low.

In terms of the survey evidence (Table 7.3), 72 firms (41 per cent) of the total sample claimed a part-time research and development effort. As might be expected, the majority of these firms were in the smaller size categories. Businesses employing less than 50 workers accounted for 63 of the 72 firms comprising the sub-sample. The level of technical sophistication in the products of the survey firms also influenced the level of research and development commitment. Although all the firms in the survey were operating in high technology sectors, firms were further categorised into three high, medium and low technology groupings based on the sophistication of their main product (for a full description of this variable see Appendix 2). Only 12 (17 per cent) of the 72 firms with part-time research and development were classified as 'high technology'. Thus, in keeping with the generally modest character of this research approach, firms with part-time research and development tended to be mainly small and in lower technology areas of production. On a regional level, part-time research and development was least prevalent in Scotland where there was a greater tendency either to maintain full-time research and development (Table 7.3) or to have none at all (Table 7.1). This heterogeneous pattern of results for Scotland will be discussed in detail in Section 7.2.2. Over half of the sample firms in the South East of England performed research and development on a part-time basis, while the level for the sample firms of the San Francisco Bay area fell between that of the two British regions.

Perhaps a further measure of the technological competence of part-time research and development effort is provided by the propensity for the major share of the development work to be performed by the firm's owner, managing director or president. For on balance, it must be helpful to the general innovation performance of the small firm if the key executive is technically competent. Significantly, Bay area firms with part-time research and development showed a marked propensity for the key figure in the effort to be the proprietor or president. A major 19 (76 per cent) of the 26 Bay area firms with part-time research and development acknowledged their predominant part-time researcher to be the president or managing director. This compared with 14 such firms (45 per cent) for the South East and 7 (47 per cent) for Scotland.

Table 7.3 Level of internal research and development commitment by region

R & D effort	Scotland		South-East England		San Francisco Bay area	
(N = 148)	N	(%)	N	(%)	N	(%)
Full-time	27	(64.3)	20	(39.2)	29	(52.7)
Part-time	15	(35.7)	31	(60.8)	26	(47.3)
Total	42	(100.0)	51	(100.0)	55	(100.0)

Chi-square = 5.93 $\qquad$ $p = 0.051$

Moreover, evidence on the proportion of work effort devoted to research and development by the predominant part-time innovator showed the Bay area executives to be more active. For example, 62 per cent of the Bay area executives with the major responsibility for research and development spent over 25 per cent of their annual work time on development compared with 33 per cent of such executives in the South East and 36 per cent in Scotland.

7.2.2 Full-time research and development

Since full-time research and development, due to its relatively high cost, is rarely undertaken in a spendthrift manner, its presence within firms implies a high level and quality of internally derived technical information. Indeed, the medium- to long-term benefits of research workers must be evident to ensure their survival (Oakey 1981). The transition from an absence of research and development effort, through part-time status, to the establishment of a full-time research and development staff is clearly a general reflection of firm size. With increasing firm size the owner/key innovator role becomes impractical for the founder due to the demands of other administrative areas in the expanding business. This problem, plus the need for innovation, is ameliorated by the positive financial benefits of expansion which, through profits, allow the employment of full-time research and development staff. This organisational change relieves pressure on the founder while still allowing for his technical input to the full-time research, if required.

Table 7.3 has indicated the high level of full-time research and development in Scotland within the sub-sample of plants with internal research and development. Indeed, even after allowing for its higher proportion of plants with no research and development facility, Scotland recorded the highest regional proportion of plants with full-time research and development expressed as a proportion of the *total* Scottish sample, with a 50 per cent level as compared with 33 per cent and 48 per cent for the South East of England and the San Francisco Bay area respectively. Beyond the initial striking observation that the incidence of full-time research and development in Scotland is marginally higher than in the San Francisco Bay area, and taking 1970 as a breakpoint, only 13 of the 29 American firms with full-time research and development were established after 1970 compared with 22 of the 27 firms in the Scottish development region (Table 7.4). Moreover, 14 of the 27 Scottish firms with full-time research and development were classified as high technology (see Appendix 2). The poor performance of the South East of England in terms of full-time research and development is partly vindicated by the region's good overall performance in full- and part-time research and development combined. Generally, however, the South East's performance is akin to that of the Bay area in that Table 7.4 indicates that the relationship of firm age to full-time research and development facilities in the South East shows a majority of firms in the pre-1970 category, while in the Scottish instance the majority is clearly in the post-1970 grouping. Indeed, a significant minority of mainly high technology newer firms have produced a heterogeneous picture of innovation in Scotland, where a particularly innovative sub-sample of firms masks the performance of a notably backward group of firms to produce the indifferent overall level of regional innovation apparent in Chapter 5.

Table 7.4 Firms with full-time research and development by age and region

Age	Scotland		South-East England		San Francisco Bay area	
(N = 76)	N	(%)	N	(%)	N	(%)
Pre-1970	5	(19.0)	14	(70.0)	16	(55.0)
Post-1970	22	(81.0)	6	(30.0)	13	(45.0)
Total	27	(100.0)	20	(100.0)	29	(100.0)

Chi-square = 13.85 $p = 0.005$

However, the mere presence of a full-time research and development facility might be a misleading measure of research and development effort. Various measures of the degree of commitment to full-time research were available to this study, but financial measures such as the research and development expenditure per worker in survey firms, which controlled for firm size, did not standardise for the higher wages paid in the United States for any given level of expertise. Since a major part of research and development costs in small firms is wages, a comparison on this basis would be invalid. A constant measure of research and development effort that again controls for firm size, but does not suffer the disadvantages of capital outlay, is the number of research and development workers expressed as a proportion of the total workforce. Table 7.5, which utilises this measure, confirms that not only does the Scottish region possess a significant minority of firms with full-time research and development, but that the levels of full-time commitment for any given percentage employment size band compare very favourably with the Bay area.

Table 7.5 Percentage of total workforce in research and development by region

R & D workers as a percentage of total workforce	Scotland		South-East England		San Francisco Bay area	
(N = 76)	N	(%)	N	(%)	N	(%)
1–9	10	(37.0)	11	(55.0)	11	(37.9)
10–24	13	(48.1)	7	(35.0)	13	(44.8)
25–49	2	(7.4)	2	(10.0)	2	(6.9)
50 and over	2	(7.4)	0	(0.0)	3	(10.3)
Total	27	(100.0)	20	(100.0)	29	(100.0)

7.2.3 *The monitoring of research and development objectives and costs*

As with other functions of the small firm, the precision of research and development objectives and of the cost-benefit analysis of their achievement is strongly related to the general sophistication of the business. To many executives, research and development is a response to perceived product specification needs as and when they occur. Because such work tends to be both intermittent and performed in spare time, questions on the broad objectives and costs of research and development often embarrass respondents because they are forced to think about a frequently expensive part of their firm's operation for which they know they should have precise answers, when in fact they have often not considered the subject in a formal manner.

Table 7.6 Main objective of research and development effort by region

R & D motive	Scotland		South-East England		San Francisco Bay area	
(N = 147)	N	(%)	N	(%)	N	(%)
Alleviate technical bottlenecks in production	3	(7.1)	3	(6.0)	2	(3.6)
Incremental product and process improvement	14	(33.3)	17	(34.0)	18	(32.7)
New product development	25	(59.5)	30	(60.0)	35	(63.6)
Total	42	(100.0)	50	(100.0)	55	(100.0)

Survey evidence on this topic in Table 7.6 confirms that a relatively consistent two-thirds majority of the sample firms acknowledged the development of new products as the main objective of their research and development effort with a further third mentioning incremental improvements in product design as the main objective, and with only a small minority in all regions stating that their research and development effort was directed merely at alleviating technical bottlenecks in production. Hence, overall, the 90 firms claiming that their research and development was directed at new product design equals 52 per cent of the survey total. Although directly comparable data for other industries are not available, the survey firms' level of commitment to new product development is probably comparatively high and reflects the generally high technology nature of these firms. These results on research and development motives had no particular regional bias.

However, if respondents were reasonably clear about their firm's research and development objectives, they were generally less precise about expenditure on research and development. In Britain, of the Scottish and South-Eastern firms with research and development effort, only 50 per cent and 47 per cent respectively attempted to calculate their research and development expenditure. The higher 73 per cent level of estimation in the San Francisco Bay area may reflect a more sophisticated approach to research and development costing in particular and to efficiency in general. None the less, the significant lack of knowledge of research and development expenditure, particularly in Britain, is irrefutable proof that not even the crudest form of cost-benefit analysis is undertaken in many of the survey firms. These results are further indirect proof of the informality of research and development in small firms.

Regional variations were detectable in responses to a question on whether firms attempted to calculate their return on research and development expenditure. Only 17 per cent of Scottish firms acknowledged this practice, while the figures for the South East and Bay area were at the higher levels of 35 per cent and 32 per cent respectively. Since the practice of research and development cost-benefit evaluation must, at least in part, be associated with sophistication, such results may indicate a weakness in Scottish firms, although this result is not supported by any of the other survey data gathered on those Scottish firms with research and development facilities. Again, given the general high technology nature of firms in the survey sample, the level of research and development cost-benefit monitoring is not impressive and underlines the informal nature of much research and development in small firms.

Interestingly, the overwhelming method of ensuring that costs did not exceed benefits in the 67 per cent of survey firms with internal research and development was a rather crude incorporation of development expenditure on a given product in subsequent prices. Indeed, in many instances, the final price of the product resulting from such an exercise inevitably involved the spreading of research and development costs over the total estimated number of units to be sold, with a healthy profit margin to allow a margin of error on estimated sales. In reality, this approach is not a true evaluation of the benefits of research and development, but merely a means of ensuring that direct capital outlay on research and development is recovered. Such a method ensures that the customer inevitably pays for the cost of research and development in the price of the product. In instances where the development costs are extensive and the units produced are few (perhaps even one), research and development costs may be the bulk of total costs. High technology production frequently displays this type of producer-consumer price relationship because the producer is technologically dominant and is able to put a premium on the price of a product that has a high technological specification.

7.3 External technical information and internal innovation

It is clear from the preceding analysis that the majority of the high technology small survey firms maintain some form of internal research and development effort. Consequently, the general value of externally acquired technical information must be largely of a supportive nature in enhancing the overall internal research and development effort of the firm. None the less, it is also true that an externally available source of technical information such as a university might give a significant boost to the internal research and development effort of small high technology firms. Indeed, it has been argued by writers in the United States that the concentration of high technology industry in both the Route 128 area near Boston and the Silicon Valley complex south of San Francisco owes much to the stimulation of technically oriented local universities (Deutermann 1966; Gibson 1970; Cooper 1970).

However, much of the reasoning on the links between technical universities and local industrial growth depends on concomitance rather than on direct causality. Certainly there is little hard evidence on locationally meaningful *information flows* between universities and local firms. It is often assumed that, because local firms with technical needs and universities with relevant expertise are closely juxta-positioned, interaction occurs. However, detailed evidence on the extent of links between high technology scientific instruments firms and universities in Britain indicated that contacts were infrequent and of a low technology supportive nature when they occurred (Oakey 1979c). Clearly, the current survey is an ideal opportunity to shed some light on possible international differences in the technical links of small firms with external research institutions.

Survey firms executives were asked if their firms maintained any external contact with a source of technical information of importance in developing products and processes in the plant. Surprisingly, firms in the San Francisco Bay area recorded the lowest level of external research and development links, with only 14 firms (23 per cent) acknowledging a link compared with approximately half of the sub-samples in Scotland and the South East of England (Table 7.7). There was evidence to indicate that local universities played an important part in producing the higher level of contact in Scotland. However, there appears to be little correlation between the most innovative region (the Bay area) and extensive local and national links with external research institutions. Moreover, an attempt was made to ascertain the strength of the information flow in the minority of survey firms with a technical link to an outside body. Executives whose firms had such links were asked if the breaking of the information flow, for whatever reason, would cause serious disruption to their internal innovation performance. A very small minority of only 6 Scottish, 4 South-Eastern and 3 Bay area firms responded affirmatively to this question.

Table 7.7 The incidence of important technical information contacts by region

Technical contacts	Scotland		South-East England		San Francisco Bay area	
(N = 173)	N	(%)	N	(%)	N	(%)
Important contacts	30	(55.6)	25	(42.4)	14	(23.3)
No important contacts	24	(44.4)	34	(57.6)	46	(76.7)
Total	54	(100.0)	59	(100.0)	60	(100.0)

Chi-square = 12.54 $p = 0.002$

These data give a clear overall impression that external technical information links are not profuse and that those that do exist are not of great significance to innovation in the survey firms. The impression created by the previous literature (Deutermann 1966; Cooper 1970) that American firms might maintain more abundant and technically important links with local and national technical universities is found to be false in the Bay area sub-sample. Indeed, this result is particularly striking given that the majority of Bay area firms were located in Silicon Valley, no more than 10 kilometres from Stanford University.

7.4 Research and development, an in-house imperative

It is clear from the preceding evidence that there is a strong link between research and development and the creation of new products in high technology small firms. Such a strong link is not only reflected in this hard evidence, but also in the opinions of many respondent executives as they described their fight to maintain or increase research and development efforts in the face of ever-present financial constraints. In most cases, the firms with no research and development are those with no management or technical expertise with which surplus internally generated capital may be turned into better products through in-house development efforts. In specific instances, certain low technology jobbing subcontract firms, that had accumulated substantial profits without a recognisable in-house product, were at a loss to decide how to invest this capital in product development. They had no product ideas, nor the ability or confidence to perform the necessary preliminary research and development. However, there was little doubt among such executives that research and development, had it been possible, was a worthy area of investment. Research and development is universally recognised by small firm management as a key to substantial small firm progress through product innovation.

There is no doubt, however, that research and development is a very costly exercise with unpredictable benefits. Problems of design and construction are great consumers of man-hours and it is impossible to 'cost' the effort required to perfect a new product embodying new technologies. Indeed, firms working for government defence departments on security-related products are often given substantial funds for development in advance of any workable prototype. These financial inducements are essential due to the 'uneconomic' nature of much research and development in areas of technology where prohibitively expensive 'trial and error' is common. However, such government practices mainly benefit large corporations.

Nevertheless, despite the problem of research and development costs, high technology firms remain viable because they pass the high cost of research and development on to their customers through the eventual price of the finished product. Critically, the reason why it is possible for high technology firms to pass on high research and development costs to their consumers is the relative uniqueness of their product specification, ensured by the research and development effort. Consumer firms are prepared to pay high prices for the high technology products essential to their own manufacturing process and for which they do not have the in-house technical expertise. This line of argument evokes a re-emergence of the 'technical dominance and dependency' concept proposed in a linkage context in Chapter 6. Indeed, much of the value added as input linkages are transformed into outputs is attributable to the extra premium that research and development embodies in the products of high technology firms.

The small firm environment is a particularly fertile basis for research and development and subsequent innovation. While quantitative comparisons between innovation levels in large and small corporations are somewhat spurious, due to the different nature of research and development in these disparate types of production unit, there is now ample evidence to indicate that many small firms are particularly innovative (Bolton 1971; Von Hippel 1977; Townsend *et al.* 1981; Rothwell and Zegveld 1982). Certainly it is clear that large American corporations do not doubt the innovation potential of small firms since they are prepared to invest heavily in new small firm ventures (Von Hippel 1977; Roberts 1977). The strong internal bond in small firms between research and development effort, management, and shopfloor personnel, and their combined ability through a small level of operation to interact on an informal basis, frequently facilitate rapid technological progress and stimulate a constructive teamwork atmosphere. In such an intense environment, embryonic ideas may be incubated to maturity with little of the bureaucratic friction apparent in larger organisations.

Perhaps because such a strong internal bond exists in most successful high technology firms, there is little scope for more tenuous links with external bodies. It is certainly true that the survey firms described in this chapter did not indicate any significant evidence of external information links in general, nor of important local innovation influencing contacts in particular. It was striking that in the San Francisco Bay area, expected links between firms and local universities, particularly Stanford University, were not important. In the absence of such links, it is more likely that the stimulative influence of Stanford University on high technology small firms in the Bay area is as a provider of skilled research and development manpower and new graduate and professorial entrepreneurs (Zegveld and Prakke 1978; Bullock 1983). But, once these personnel leave the local university to establish or become employed in local small firms, technical contacts seem to be infrequent and unimportant.

8 Labour and innovation

8.1 Labour and location

8.1.1 *Labour and innovation in high technology industries*

In high technology industries, with their strong emphasis on value added in production through human research and development and production skills, it seems reasonable to assume that labour will be of great importance. Moreover, it might also be argued that shortages in local labour skills could potentially stifle the innovation performance of the high technology small firms of this study. However, unlike large corporations in which manufacturing plants may be established at short notice, leading to a large and sudden short-term demand on the local labour market (Townroe 1975), the employment needs of most small high technology firms tend to be long term and incremental in nature. Typically, a small firm is established in the area of the founder's home, with small premises and a commensurately small labour force — perhaps five workers including the founder (Oakey 1981). In the early stages the labour demands of the firm are dictated by profitability. Hopefully, growing demand and profits prompt the employment of additional workers as and when they are required. This incremental approach to labour demand is advantageous in that the small firm, unlike large corporations, rarely puts any substantial strain on the local labour market and thus may more flexibly vary the timing of employment increases in tune with local labour availability.

These arguments tend to suggest that labour shortages in the small high technology firm, although potentially damaging to innovation, are more inhibitive than prohibitive in their effect. Unlike the large corporation, which may reject a particular region as a potential production location due to a poor or congested local labour market, most small independent firms already exist in the region of their birth and are resolved to grapple with the problems of their own particular labour market, partly because they know no other (Oakey 1981). But notwithstanding such contextual reservations, shortages of key workers in firms that are rapidly expanding may cause considerable friction. In such conditions the small high technology firm faces greater problems than its low technology counterpart, both because of the relative speed with which expansion occurs, and because of the higher quality and broader spectrum of skills required. Chapter 7 has confirmed that most small high technology firms perform internal research and development, which implies a general need for skilled research and development manpower. However, since both the essential development *and* the production phases of manufacture in high technology small firms are generally skilled-labour-intensive, there is reduced scope for the employment of semi-skilled or unskilled workers, which generally complicates the employment process. For example, these characteristics contrast sharply with those of a low technology small subcontract firm supplying the engineering industry in which research and development is not necessary and semi-skilled machinists might be sufficient to perform shopfloor tasks.

However, since the labour requirements of high technology firms may be both fast-growing and exacting, it might be argued that shortages of labour in a region will be influenced more by changes in demand than in supply, particularly in regions with many expanding firms in a spatially constrained industrial agglomeration. Indeed, it is clear that in extreme cases, small firm innovation may be inhibited by labour shortages. However, it is ironic that this inhibition is likely to cause performance reduction in the *most innovative* expanding businesses, since these firms put the heaviest pressure on local labour market resources.

8.1.2 Supply and demand factors

Traditional economists view the operation of wage levels in national labour markets in terms of equilibrium theory in which short-term fluctuations above and below national average wage levels in particular regions will adjust themselves in the medium to long term (Gitlow 1954). The nub of this argument is that shortages of labour in a given labour market will cause wages to rise, which will in turn attract migrants from other lower-wage labour markets, thus producing a matching of supply to demand and a subsequent fall in wage levels in the high-wage labour market toward the national average, thus causing equilibrium. However, it is clear that the credibility of such a theory largely depends on the medium-term mobility of labour and its reflexive response to, implying knowledge of, high potential wage levels in other regions. Certainly, in the British instance, shopfloor labour in the high technology instruments industry has been noted to be particularly immobile and a major constraint on the mobility of instruments industry production (Oakey 1979b).

If individuals in economies with lower wage levels are reluctant to move in adequate numbers to areas of greater wage advantage, it is likely that the potential host region will 'overheat' in terms of wage levels and that equilibrium will not ensue: indeed, the problem will probably worsen. Moreover, if the complex problems of high technology industry are introduced into this discussion, two other major factors confuse the issue. First, equilibrium theory seems to imply that growth in the prosperous area stabilises at some high equilibrium to allow the influx of high-wage-seeking labour to satisfy the demands of production, and thus cause a fall in wage levels. However, evidence from high technology industrial agglomerations such as Silicon Valley tends to suggest that growth begets more growth and that the influx of labour does not keep pace with the rising demand, thus having little effect on wage levels. Second, the theory is deficient in that it seems to imply that a car mechanic in Detroit or Manchester can become, virtually 'overnight', a skilled instrument technician in Sunnyvale or Basingstoke. While highly paid jobs may exist in high technology agglomerations, suitably qualified personnel are frequently not available in the short to medium term, either in the agglomeration or elsewhere. In such circumstances it is perfectly feasible that, in the teeth of a recession in both Britain and America, there are a substantial proportion of high technology firms experiencing labour shortages. Ironically, the equilibrium theory *is* accurate in so far as it initially anticipates local wage inflation in response to skilled labour shortage.

The shortage of labour in any given local economy is an effect that ensues from one of two major causes. First, taking the example of a depressed region, labour shortages can occur where the supply of a particular type of worker is low. In those high technology industries that do exist in such regions, shortage of skilled workers may cause labour problems, predominantly because a moderate demand is not met by an inferior supply. Often this problem confronts large foreign firms seeking to locate a branch plant in one of Britain's development regions. Since training is not practicable in the short term, the immediate dearth of suitable high technology skills in the depressed region repels the potential locator, thus helping to perpetuate an inferior regional skill structure. Second, at the other extreme, labour shortages may occur in areas where high technology industry is common in circumstances where a large, skilled, high technology labour market acts as a retentive force for existing firms and an attractive force for other companies from outside the agglomeration.

In these circumstances, labour shortages are caused by a large local skilled workforce being unable to satisfy the greater demand of local firms. While it is true that firms may be attracted to this agglomeration because of its large 'pool' of skilled labour, it is critical to note that virtually all this labour 'pool' is *employed* at any given moment in time. The manner in which firms acquire or increase their workforces in such an area is by outbidding their fellow firms, thus causing escalating wage levels and higher production costs. However, these high production costs are economically sustainable because of the high prices that high technology products can command. The high wage levels paid by high technology firms in such agglomerations are indirect proof of the essential nature of these workers, not only to ensure continued growth but also, in many instances, to allow the continued functioning of the firm.

Thus, for totally different reasons, skilled labour shortages can be as acute in areas of high technology industrial employment concentration as in depressed areas with impoverished skill structures. It is indeed a paradox that the greatest shortages of highly skilled labour can in many instances occur in areas where the concentration of such labour is particularly high. If the additional problems of high wage levels and increased living costs, notably housing, are added to the problem of labour shortages in areas with concentrations of high technology workers, such areas might initially appear singularly unattractive to high technology firms. But it is again paradoxical that the drawbacks of location within a high technology agglomeration with all its attendant labour problems are, in a sense, evidence of the area's retentive powers. Indeed, as mentioned previously, problems with labour supply in such innovative areas are frequently evidence of the effects of rapid innovation-led expansion, partly spurred on by the many other clear advantages of the location. Seen in this light, problems of labour shortage or cost may be a lesser evil when compared with the converse problems of lay-offs, redundancies and other manifestations of small firm decline and death, often associated with areas where labour is cheaper in depressed economies.

In the remainder of this chapter these assertions on labour and high technology small firms are related to evidence on the impact of varying regional labour supplies on innovation in the survey firms. However, the direct measurement of problems associated with labour shortage are preceded by introductory contextual evidence on any regional variations apparent in the general skill levels in survey firms.

8.2 Regional variations in the proportion of shopfloor workers in survey firms

As an initial step in this investigation of the effects of labour supply on innovation in small high technology firms, it is valuable to examine variations in the proportion of shopfloor workers employed in the survey firms. The number of shopfloor workers expressed as a percentage of the total workforce in order to control for variations in firm size, has in the past proved to be a generally reliable indicator of a firm's commitment to innovation. For example, firms with a high proportion of indirect workers will mainly owe this particular characteristic to a large proportion of executive and research and development workers, reflecting greater overall technical sophistication of these firms (Thwaites *et al.* 1981). To test this asserted link between the level of indirect workers and technical sophistication, the proportion of indirect workers in the survey firms was compared with the previously used technological complexity variable, which is a further division of these generally high technology firms into high, medium and low technology categories based on the sophistication of the main output (see Appendix 2 for a fuller explanation of this variable). It is certainly true that, taken at the level of the total survey sample, there is a strong, statistically significant propensity for the highest technology firms to record a low proportion of shopfloor workers (Table 8.1). The chi-square test on this tabulation easily exceeded the $p = 0.05$ level of significance.

Table 8.1 Technological complexity by shopfloor workers as a percentage of total workforce

Shopfloor workers as a percentage of total workforce	Technological complexity					
	High		Medium		Low	
(N = 174)	N	(%)	N	(%)	N	(%)
1–24	5	(10.0)	4	(7.7)	2	(2.8)
25–49	18	(36.0)	17	(32.7)	7	(9.7)
50–74	21	(42.0)	19	(36.5)	30	(41.7)
75–100	6	(12.0)	12	(23.1)	33	(45.8)
Total	50	(100.0)	52	(100.0)	72	(100.0)

Chi-square = 26.10 $p = 0.0002$

However, the regional evidence on the proportion of shopfloor workers in the survey firms is not straightforward. While Scotland and the San Francisco Bay area perform similarly in that 35 firms in both regions devoted over 50 per cent of their workforce to shopfloor tasks (i.e. 65 per cent and 58 per cent respectively), the same category in the South East of England comprised 51 businesses (85 per cent) (Table 8.2). Taking the results of Tables 8.1 and 8.2 together, it might initially be argued that these 51 South Eastern firms were predominantly lower technology businesses, since Table 8.1 has indicated that there is an overall link between low technology status and a high incidence of shopfloor workers in survey firms. On reflection, however, this possibility is rendered unlikely by the general observation that low technology firms are not particularly prevalent in the South East. Indeed, the regional consistency of the survey total of 72 low technology firms is strikingly consistant with 24 firms (33 per cent) in Scotland and 26 (36 per cent) and 22 (31 per cent) businesses in the South East and Bay area regions respectively. The South-Eastern firms with over 50 per cent of their workforce on the shopfloor must clearly include a proportion of medium and high technology firms, since the 26 South-Eastern low technology firms would not account for the 51 firms in Table 8.2, even if they were all included.

Table 8.2 The regional distribution of shopfloor workers expressed as a percentage of total workforce

Shopfloor workers as a percentage of total workforce	Scotland		South-East England		San Francisco Bay area	
(N = 174)	N	(%)	N	(%)	N	(%)
1–24	4	(7.4)	1	(1.7)	6	(10.0)
25–49	15	(27.8)	8	(13.3)	19	(31.7)
50–74	21	(38.9)	29	(48.3)	20	(33.3)
75–100	14	(25.9)	22	(36.7)	15	(25.0)
Total	54	(100.0)	60	(100.0)	60	(100.0)

Chi-square = 11.46 $\qquad$ $p = 0.075$

These suppositions are confirmed in Table 8.3, which indicates that low technology firms in the South East, although accounting for 47 per cent of the 51-total South-Eastern firms with over 50 per cent of their workforce on the shopfloor, constituted the lowest proportion of low

Table 8.3 The regional contribution of low technology firms to the sub-sample of firms with more than 50 per cent of their workers on the shopfloor

	Scotland	South-East England	San Francisco Bay area
A Firms with > 50% of workers on shopfloor	35	51	35
B Low technology firms with > 50% of workers on shopfloor	19	24	20
B as a percentage of A	54	47	57

technology firms in this category when compared with the other regions. Moreover, high and medium technology firms with over 50 per cent of their workforces on the shopfloor comprised a greater proportion of the *total* regional samples in the South East (i.e. 27 firms, 45 per cent) than in Scotland (16 firms, 30 per cent) or the Bay area (15 firms, 25 per cent). These data complement the previous results of Chapter 7 which revealed that firms in the South East, although displaying a similar level of product innovation, and with a proportionate share of high technology firms compared to the other regions, evidenced a much higher level of part-time research and development and a conversely lower level of full-time research and development employment. It is likely, then, that the higher incidence of shopfloor workers evidenced in South-Eastern firms is a corollary of this research and development evidence since, clearly, full-time research and development workers are a major component of non-shopfloor employment. It remains for subsequent analysis to discover whether these regional differences in the proportion of shopfloor workers affect the shortages of skill types in the survey firms.

8.3 Labour shortages and innovation

Bearing in mind the preceding evidence on regional variations in the proportion of shopfloor workers in the survey firms and the introductory comments on conditions that cause labour shortages, respondents were asked if they had recently experienced shortages of any particular type of labour in the local area. It is generally evident from Table 8.4 that firms in the more prosperous and innovative regions of the South East of England and the San Francisco Bay area, as noted in Chapter 5, displayed a greater propensity to acknowledge labour shortages, particularly those firms in the South East. A chi-square test yielded a significance level of $p = 0.11$, marginally outside the normally accepted $p = 0.05$ level of significance. It is notable that firms in Scotland, the least innovative region, recorded the lowest level of labour shortages, thus tending to support the introductory argument that labour shortage might be caused by acute demand outstripping a substantial supply in areas where high technology industry is concentrated. Given the earlier evidence (in Chapter 5) on the lower product innovation performance of Scotland, it is doubtful whether the lower level of problems with labour supply in this region is caused by a particularly comprehensive labour supply meeting vigorous demand. It is more likely that a poor labour supply remains adequate to suffice the lower level of demand from high technology firms in Scotland.

Table 8.4 The regional distribution of firms experiencing labour shortages

Labour shortage	Scotland		South-East England		San Francisco Bay area	
(N = 174)	N	(%)	N	(%)	N	(%)
Shortage	20	(37.0)	34	(56.7)	29	(48.3)
None	34	(63.0)	26	(43.3)	31	(51.7)
Total	54	(100.0)	60	(100.0)	60	(100.0)

Chi-square = 4.40 $p = 0.111$

However, a closer examination of the types of skill shortages acknowledged by firms with labour problems yielded detailed evidence of further interest. Interestingly, there was a consistent difference in the types of labour shortage in Scotland compared with the remaining regions. While numbers in individual cells are small, it is clear that research and development personnel constituted the main category of shortage in Scotland, while shopfloor skilled and semi-skilled workers combined were overwhelmingly the cause of shortages in the South East and the Bay area (Table. 8.5). The emphasis on shopfloor workers in the South East might reflect the earlier evidence on the proportion of shopfloor workers in the total workforce, which showed this category of personnel to be particularly high in the South East. It is probable, therefore, that the need to maintain a high proportion of shopfloor workers is reflected in the problems experienced in obtaining shopfloor workers apparent in Table 8.5. The significant regional difference between Scotland and the other regions in particular was confirmed by a strongly significant chi-square test which easily exceeded the $p = 0.05$ level.

Table 8.5 The regional distribution of skill shortages

Shortage type	Scotland		South-East England		San Francisco Bay area	
(N = 80)	N	(%)	N	(%)	N	(%)
R & D worker	8	(40.0)	2	(6.1)	3	(11.1)
Skilled shopfloor worker	5	(25.0)	17	(51.5)	17	(63.0)
Shopfloor worker	3	(15.0)	11	(33.3)	4	(14.8)
Other	4	(20.0)	3	(9.1)	3	(11.1)
Total	20	(100.0)	33	(100.0)	27	(100.0)

Chi-square = 17.15 $p = 0.009$

Firm executives acknowledging shortages of labour were questioned further in order to pinpoint specifically whether such differences had inhibited their efforts towards product innovation (Table 8.6). The results of Table 8.6 may be seen as a further development of the trend hinted at in Table 8.4. Again, it is clear that, as might be expected from the preceding results, labour shortage is less of an inhibitor of product innovation in the Scottish instance. The Scottish region, which possessed the lowest proportion of firms with labour shortages also recorded the lowest proportion of firms acknowledging that labour shortages inhibited product innovation, with a 10 per cent level of affirmative responses among those firms with labour problems as compared with a larger 41 per cent and 48 per cent level among such firms in the South East and Bay area respectively (Table 8.6). These sharp differences between the three regions are confirmed by a chi-square test, significant at the $p = 0.02$ level. This result reflects the increased regional difference between labour shortage *per se*, in which a weaker significance of $p = 0.111$ was achieved (Table 8.4), and the current direct investigation of the impact of such shortages on innovation in the survey firms. Moreover, although cell numbers become particularly small, a breakdown into skill types of those labour shortages which, in Table 8.6, firms acknowledged as inhibiting innovation, concurs with Table 8.5 by indicating that the 2 Scottish firms in Table 8.6 both acknowledged research and development personnel as the skill type whose shortage inhibited innovation, while in the South East 13 of the 14 firms mentioned shopfloor workers and in the Bay area 9 of the 14 firms in Table 8.6. also mentioned shopfloor personnel.

Table 8.6 The regional distribution of firms in which labour shortages inhibited innovation

Effect on innovation (N = 83)	Scotland		South-East England		San Francisco Bay area	
	N	(%)	N	(%)	N	(%)
Inhibition	2	(10.0)	14	(41.0)	14	(48.0)
None	18	(90.0)	20	(59.0)	15	(52.0)
Total	20	(100.0)	34	(100.0)	29	(100.0)

Chi-square = 8.16 $p = 0.020$

The results thus far combine to create an overall impression that generally suggests that the problems of labour supply are most acute in the more prosperous regions. However, it is fair to state that the inhibiting influence of labour shortages on innovation is marginal, since of the total regional samples, a minority of 4 per cent of Scottish firms, 23 per cent of South-Eastern firms and 23 per cent of Bay area firms stated that labour shortages directly inhibited their innovation performance. None the less, in individual instances the debilitating effects of labour shortage can be intense. The noted importance of shopfloor workers in areas of high technology industrial concentration confirms earlier results on the British instruments industry (Oakey 1981).

8.4 Acquisition behaviour

Given that all firms, whether they experience problems of labour shortage or not, will need additional or replacement workers from time to time, methods of recruitment may shed further light on the acquisition process and how it relates to success or failure in the job market. Clearly, high technology industries, with a uniformly high general level of skilled specialist workers, might be expected to maintain training programmes to create skilled workers.

8.4.1 In-house training

Although expensive, training programmes offer the valuable advantage of fitting the trainee to the precise needs of the firm. Conversely, in a time of deep recession in both Britain and the United States, skill training is a 'luxury' many small businessmen consider they are no longer able to afford. Table 8.7 indicates the regional extent of training programmes in survey firms. While the British regions perform very similarly, with levels of training programme acknowledgement marginally above 50 per cent, firms in the San Francisco Bay area show a noticeably greater tendency for training programmes, with a 70 per cent level of acknowledgement. In view of the preceding results on the predominant shortage of blue-collar workers, and of the evidence from many conversations with individual executives, it was not surprising to discover that most of the training schemes mentioned in Table 8.7 are directed at creating shopfloor workers. The higher incidence of training programmes in the Bay area probably reflects the greater competition for skilled shopfloor labour rather than its absolute shortage.

Table 8.7 The regional distribution of firms operating training schemes

Training scheme	Scotland		South-East England		San Francisco Bay area	
(N = 174)	N	(%)	N	(%)	N	(%)
Internal training scheme	29	(53.7)	33	(55.0)	42	(70.0)
None	25	(46.3)	27	(45.0)	18	(30.0)
Total	54	(100.0)	60	(100.0)	60	(100.0)

Chi-square = 4.01 $p = 0.135$

8.4.2 External acquisition

Another means of creating skilled or trainable labour, both in firms with or without internal training programme, is external acquisition. While a limited number of trainees may be acquired direct from school, the main means by which experienced labour is obtained is through 'poaching' in various ways from other local firms. The following evidence is divided into shopfloor worker and executive/ research and development personnel acquisition behaviour. It was anticipated that the recruitment of white-collar workers would involve the utilisation of a spatially larger and more varied labour market than for blue-collar staff.

Table 8.8 initially views the methods used by the survey firms to obtain shopfloor workers. It is clear that two major approaches to acquisition are common in all regions for this type of worker. While the recruitment of labour through local newspaper advertisements is a covert means of obtaining workers *already* gainfully employed, the acquisition of workers through 'local contacts' is a far more direct and overt activity which often involves 'head-hunting'. This overt approach offers the clear advantage of a known personal and professional track record, and clearly, a potential employee is not considered unless he meets the firm's requirements. Government agencies are a source of labour in both Britain and the United States. However, the skill levels of such workers tend to be low, since if they had adequate skills they would probably not have needed to resort to these agencies. It is clear from Table 8.8 that such agencies were more heavily utilised in Britain, which contrasts with the importance of local contacts and newspaper advertisements in the San Francisco Bay area.

In examining behaviour towards the appointment of white-collar

Table 8.8 The regional distribution of the means by which shopfloor workers were acquired

Means of acquisition	Scotland		South-East England		San Francisco Bay area	
(N = 169)	N	(%)	N	(%)	N	(%)
Local contacts	14	(28.0)	8	(13.6)	19	(31.7)
Local Press	19	(38.0)	29	(49.2)	31	(51.7)
Trade journal	0	(0.0)	0	(0.0)	1	(1.7)
Government employment agency	17	(34.0)	17	(28.8)	8	(13.3)
Other	0	(0.0)	5	(8.5)	1	(1.7)
Total	50	(100.0)	59	(100.0)	60	(100.0)

executive and research and development personnel in the small survey firms, it is important to note that not all firm owners considered themselves in a position to justify the appointment of any additional executive or research and development workers. Thus it is initially enlightening to note as a measure of organisational sophistication the higher contribution of Bay area firms to Table 8.9, which records the means by which white-collar workers were acquired. Although firm sizes in the three regions were broadly comparable, 70 per cent of the Bay area firms acknowledged a method of acquisition for white-collar workers, compared with 48 per cent and 57 per cent for the South East of England and Scotland respectively.

Table 8.9 The regional distribution of means by which research and development and executive personnel were acquired

Means of acquisition	Scotland		South-East England		San Francisco Bay area	
(N = 102)	N	(%)	N	(%)	N	(%)
Local contacts	10	(32.3)	4	(13.8)	18	(42.9)
Local Press	8	(25.8)	4	(13.8)	9	(21.4)
Trade journal	4	(12.9)	4	(13.8)	1	(2.4)
National Press	6	(19.4)	5	(17.2)	0	(0.0)
Personnel agency	2	(6.5)	3	(10.3)	13	(31.0)
Other	1	(3.2)	9	(31.0)	1	(2.4)
Total	31	(100.0)	29	(100.0)	42	(100.0)

As might be expected with regard to the acquisition of white-collar workers, local contacts were of marginally increased overall importance in this category when compared with the acquisition methods for blue-collar workers noted in Table 8.8, with Bay area firms using this method slightly more than the survey average and South-Eastern firms using it less than the other regions. Indeed, the South-Eastern sources were generally more diverse in type than in the other regions. Interestingly, the use of professional personnel agencies was second only to personal contacts as a means of acquisition in the Bay area firms, while this method was insignificant in the firms of the British regions.

Overall, there is extensive evidence in much of the preceding data that in all the regions labour, both blue- and white-collar, was drawn from the local area surrounding the plant. Most of the acquisition methods used, particularly regarding skilled personnel, involved 'poaching' by either direct local personal contacts or indirect local newspaper advertising. These external methods of employment generation complemented a significant internal effort through training programmes in a substantial number of survey firms, particularly in the Bay area sub-sample. These data support the view that it is the most prosperous regions that experience the greatest problems of labour shortage, despite their existence within concentrations of relevant labour skills. This general conclusion forms the basis for the following closing comments to this chapter on the impact of labour supply on innovation in small high technology firms.

8.5 The causal link between labour shortage and innovation

Labour is often mishandled by economists because they consider it to be a resource input to the production process in much the same manner as capital (Gitlow 1954). However, unlike inanimate material inputs, labour is much more than a resource input. Notwithstanding the purely behavioural and often sub-optimal nature of man as an actor in the optimising world of the economist, he may be cast in the roles of inventor, production worker and consumer in the wider economy. All these complex and subtle factors combine to render labour a unique input resource to the production process. Such points have clear importance to the following arguments, since aspects of both labour as a resource factor and labour as a behavioural instigator of manufacturing industry will be considered. First, labour as a resource factor is examined.

It should be re-emphasised that the central objective of this chapter was to examine labour as a resource factor and determine whether variations in its availability and quality restricted innovation such that they might explain regional variations in the innovation performance of small firms. As a means of drawing useful conclusions from the arguments of this chapter in general, and from the data in particular, it is helpful to consider the innovation process and its effect on labour demand in cause and effect terms. Figure 8.1 is based on the, now familiar, product life cycle–firm growth model of earlier chapters. If the birth and growth of a firm is closely considered it is apparent that both the birth of the firm and the subsequent product innovation during growth are marked by innovative pushes from within the firm. In the following case the product innovation push is cited, although the decision to increase output and employment in subcontracting firms without a product base might be termed innovative in a strategic sense.

This innovative push is merely a decision to produce a new product for the first time in a new firm, or is a reiterative procedure in an established business. Each innovation implies the full cycle of events shown in Figure 8.1. However, it is important to note at a conceptual level that the innovative push is cited here as the *cause* of subsequent labour demand. Figure 8.1 emphasises the point that at the end of each product life cycle firm executives, faced with declining sales, make a decision on whether to initiate a new cycle of innovation or to shed labour in a defensive strategy. It is clear that most firms do not take such a defeatist approach and attempt, conversely, to initiate a new 'innovative push', if only to maintain existing workers. However, it should also be remembered that some small firms will be sub-contractors with no in-house product, or manufacture a low technology product with a very long product life cycle, in both cases negating the need for decisions over innovative pushes, but also missing the benefit of innovation-led rapid growth.

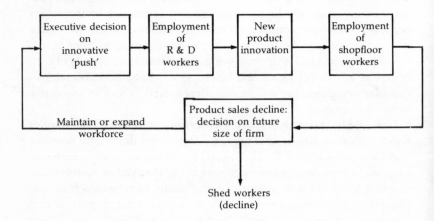

Fig 8.1 An innovation–labour demand cycle

Thus it would appear from these arguments, and from the tenor of the results of this chapter, that the main reason for the noted shortage of skilled shopfloor workers in the San Francisco Bay area in particular is vigorous innovation, since innovation must precede the growth which creates the demand for workers in small firms. A low number of skilled workers in an area is not, by itself, sufficient cause of labour shortage, since it must be accompanied by *demand*, which is largely innovation-stimulated. In the cases of the South East of England and the Bay area, areas where relevant skilled workers are known to be concentrated, the problems with labour supply must largely stem from a large reservoir of labour being insufficient to satisfy an even greater demand as hypothesised in the introduction to this chapter. Hence, while labour shortages may slow innovation in prosperous area firms, they are a sustainable cost of success when set against other advantages of high technology industrial agglomerations. For it has always been widely acknowledged that the net value of any agglomeration is not solely comprised of advantages but, more realistically, of advantages after disadvantages have been subtracted (Alonso 1971).

However, while of only contributory importance when considered as a resource input, labour might be seen as a cause of innovation if viewed in a different light. The initial innovative push in Figure 8.1 must be provided by an entrepreneur who may previously have been a small part of a labour input resource in another large firm. Clearly, if a local agglomeration is prosperous, and encompasses many high technology firms, it is likely that the entrepreneurs who set up their new firms in such an area will be of a generally high standard with a relatively aggressive attitude towards innovation. Significantly, a local agglomeration of high technology labour skills may not increase innovation in its role as a resource input, since skilled high technology labour is always in short supply in such areas, as emphasised in the preceding data. However, local innovation may be enhanced because such a skilled workforce is a rich source of new entrepreneurs (Zegveld and Prakke 1978; Bullock 1983). Given, in the San Francisco Bay area, the notable growth of individual small firms into multinational corporations in the past twenty years, it could justifiably be argued that labour, in its role as a source of new firm formation, is a potent indirect stimulator of innovation. However, as already mentioned, the accommodation of this view of the effects of labour on innovation involves a broadening of the concept of labour in industrial production beyond the input resource definition considered in this chapter. Such broader discussions of the general growth implications of skilled labour in high technology industrial agglomerations are reserved for the concluding chapter of this study.

9 Finance and innovation

Investment capital, as fuel for the innovation process, is a necessary early input to any substantial product development. However, the degree to which direct research and development expenditure is an overt problem to innovation in the individual firm will depend on how formally the research and development effort is organised, and more specifically, on whether firms maintain a full-time research and development commitment. It is clear from Chapter 7 that the formality of research and development in small high technology firms will vary, both in terms of its full- or part-time status, and with regard to its accurate financial monitoring. It is generally true that the firms with a full-time research and development effort will be less able to ignore the costs of their salary-earning research and development staff. But regardless of whether firms formally calculate the cost of their research and development effort, development expenditure will be a drain on the scarce resources of small firms and may put to the test the resolve of executives to pursue a particular project if the costs of development escalate. In such circumstances, the dichotomy between firms that deliberately evaluate and make provision for research and development funds, and those that allow a research and development project to 'evolve' on an informal basis, is often irrelevant. For the rapidly escalating costs of a development project will become readily apparent, even in firms that do not formally monitor such costs.

9.1 Finance and the innovation process

9.1.1 Innovation finance and risk

Probably the greatest misconception regarding the innovation process is the implication that research and development effort inevitably results in successful innovation. In reality, research and development can be a very expensive business, while its benefits are uncertain. As emphasised in earlier chapters, the nature of the innovation cycle ensures that research and development expenditure runs at least two to three years in advance of any return from subsequent related product sales, and is often at its height when the profits of the firm are at a low ebb (see Section 7.1). In such circumstances, small firm executives in financial difficulty may be tempted to write-off past expenditure by abandoning a promising product development with escalating research and development costs. In this context, there are two separate but cumulative forms of risk involved in the continued development of such a new product.

First, there are the mainly technical risks which accompany any manufacturing development. These are many and varied and range from the danger of being beaten to the marketplace by competitors, to simply not achieving the standard of specification necessary for the new product to function in an envisaged sales niche. While success or failure is rarely absolute, there remains great scope for the prototype that does not meet the minimum standards required for appearance in the marketplace. However, it is clear that failed products will not be widely publicised by the firm concerned, for obvious reasons of technical credibility. The financial risks involved in product development generally increase in direct proportion to the amount of time and money spent on a project. Ironically, it is most frequently the most high technology firms with the greatest innovation potential that are caught in the trap of over-extending their financial capacity through the laudable desire to develop products at the leading edge of technology.

The second set of risks that confronts the small firm executive compounds those associated with the technical viability of a new product. Despite the *eventual* technical success of a particular new product development, there remains the need to ensure the day-to-day survival of the business. In the short term, the materials for the current products that produce internal funds for research and development must be puchased. Clearly, in times of falling profit from ageing current products, cash flow problems may arise. In such circumstances, the firm is faced with a stark decision. It can either reduce running costs through economies, or it can borrow money to ease the firm through financial stress points in the innovation cycle (see Figure 7.1). While individual sources of capital are considered in detail in the following pages, it is critical to emphasise here that a new set of risks are introduced when a firm's management is forced to consider external means of supporting the internal innovation process. To begin with, inevitably high interest rates on any external loan will make an external solution more expensive and thus more risky.

This dilemma is perhaps best exemplified by a small firm owner who considers a second mortgage on his own house in order to continue a product development on which he has no technical doubts. While such an individual may have total confidence in the eventual *technical* viability of the project concerned, the potential loss of his own house, with the consequent effects on his family, may test his nerve to breaking point. In conditions where bankers and other loan agencies are sadly ignorant of the technical potential of a new product, the short-term financial weakness of the company due to research and development costs may cause the collapse of the whole company as it reaches the 'last lap' in innovation terms, just as the new product is about to come to fruition. The technical advances inherent in the new product are often evaluated by accountants who, due to the nature of their training, may be more interested in the mundane but financially quantifiable assets of the firm, such as the value of the plant or buildings, when determining the advisability of the extension of credit. Faced with such problems, many small firm executives may consider the financial risks to the firm of continuing an expensive product development prohibitive and abort the project despite their confidence in the technical merits of the new innovation. For it is clear that in extreme instances, a costly research and development project that will not bear fruit in the short term may cause the demise of the whole firm. These general assertions will be reflected in the evidence reported in the following sections.

9.1.2 *Sources of investment finance*

Given the added uncertainty associated with the acquisition of external investment capital for internal research and development projects, it should not be surprising to discover a tendency to rely on internal profits as a means of funding research and development. Much of the strategy towards external capital borrowing will be determined by the scale and sophistication of the research and development operation inherent in the firm. Clearly, smaller low technology firms will be more inclined to move forward incrementally, performing research and development on a small scale and only spending surplus cash as and when it becomes available, in a flexible manner. Such a defensive approach to innovation is understandable, particularly in view of the recent economic conditions in which interest rates on borrowing have been particularly high. However, while this approach is likely to avoid the problems of financial over-extension, already discussed in the context of risk, it will rarely yield the large profits associated with radical leaps forward at the leading edge of product technology.

It is likely that vigorous research and development investment in a small firm is less attributable to size or age but is more reliably determined by the level of technology inherent in its products. Indeed, new small firms entering high technology product markets are exposed to intense specification competition. Such competition demands that they must perform serious research and development in order to survive, since a product that still performs at 1975 tolerances will not sell in today's markets. In these rarified conditions, where the return on previous sales may not be adequate to fund current research and development or, in the case of new firms, where previous sales do not exist, ambitious executives with burgeoning technical ideas may approach external bodies for the necessary additional investment capital. Firms of the high technology type are aided in their search for external assistance in two fundamental ways.

First, because their products are high technology, banks, government agencies or private investors will be more favourably disposed to the extension of loans. This argument does not negate earlier criticisms of the lack of technical knowledge on the part of loan agencies, since their technically based predisposition towards firms in, say, the computer industry is predominantly at the layman's level of expertise, and this progressively becomes a problem as the loan application is pursued. However, regardless of motive, loan seekers from 'trendy' industries initially benefit from these general preconceptions when compared with low technology firms. Second, it is common for executives in high technology companies to have undergone a higher education. This experience renders them more likely to be able to prepare loan applications to banks and be more generally able to argue their case with professional agencies when compared with their blue-collar counterparts. It is generally true that graduates are more likely to possess the skills necessary to be successful businessmen in both technical and general entrepreneurial contexts (Johnson and Cathcart 1979).

9.1.3 *Investment finance in the survey regions*

Since the central theme of this book is the examination of the effect of local regional resources on the innovation performance of the survey firms, it might be expected that variations in the availability of local external investment capital for expanding high technology small firms might influence the extent and pace of innovation. The chosen study regions are well-suited to this task since they are individual examples of diverse financial environments, ranging from the public-sector-assisted development region of Scotland, through the British unassisted area of the South East of England, to the San Francisco Bay area of California, perhaps the modern home of venture capitalism. Apart from the obvious national diversity of assisted and non-assisted area status in Britain, the inclusion of the Bay area firms offers a much wider comparison of institutional, political and behavioural attitudes towards investment and innovation. In particular, these internationally comparative data offer the opportunity to evaluate the contribution of venture capital to the impressive small firm growth in the Silicon Valley area of California.

Bearing these introductory comments in mind, the empirical sections of this chapter appropriately begin with the problems of investment finance and innovation in new small firms. The discussion will focus on the needs of new firms and their potential, given adequate capital resources, for employment generation in high technology industrial areas.

9.2 Finance, innovation and the new firm

9.2.1 Sources of start-up capital

Earlier chapters have emphasised that the new firm is similar to the existing small firm in that, assuming the new business in question prospers, the problems it encounters at birth are repeated during growth (Chapter 7). In this sense, the initial product life cycle is distinctive only in that it may be the first of many such innovation processes during the life of the firm. However, it is clear that the initial innovation in those firms that produce a recognisable product from birth is particularly traumatic. The problem of investment finance is acute for new businesses because, by definition, they do not have access to profits from previous sales with which to aid development. Moreover, the new firm's standing with banks is exacerbated by the lack of a 'track record' with which to argue the case for a loan. The regularity with which this observation has occurred in Britain during the past fifty years suggests that the problem has not been solved (Macmillan 1931; Bolton 1971; Wilson 1979). With these initial comments in mind, it will be enlightening to discover the origins of the start-up capital of newer high-technology firms in the survey sample.

Perhaps in view of the recent high interest rates in Britain and the United States, it is not surprising to note that 62 per cent of all survey firms founded since 1970 relied on the personal savings of the founder for the initial injection of capital with which to start the business. The sums involved may be very small, and survey evidence gained from conversations with founders suggests that in many cases the business is born with the aid of an increased bank overdraft of a few hundred pounds or dollars. Indeed, the business is often begun while the founder remains employed at a large corporation, the full-time operation of the firm taking place only after a trial period in which sales and experience are accumulated. Such an approach is fundamentally incremental and, although relatively slow, offers the major advantage of reducing the risks of setting up, when compared with faster loan-based launches.

Regional evidence on the initial main source of start-up capital indicates the universal popularity of personal savings as a means of beginning a firm. The 62 per cent average level of acknowledgement for this source is similar to levels observed in other studies of small firm start-up finance (Cross 1981; Storey 1982). While a wide range of other sources are sparsely represented, the only other important origin of start-up funds is private venture capital (Table 9.1). It is also clear that venture capital's status as the second most important source of funds is caused by the San Francisco Bay area sub-sample of firms,

Table 9.1 Sources of start-up capital in firms founded since 1970 by region

Start-up capital	Scotland		South-East England		San Francisco Bay area	
(N = 73)	N	(%)	N	(%)	N	(%)
Personal savings	20	(67.0)	11	(69.0)	14	(52.0)
Previous assets	0	(0.0)	2	(12.0)	2	(7.0)
Bank loan	2	(7.0)	1	(6.0)	1	(4.0)
Second mortgage	1	(3.0)	0	(0.0)	1	(4.0)
Venture capital	1	(3.0)	1	(6.0)	8	(30.0)
Other	6	(20.0)	1	(6.0)	1	(4.0)
Total	30	(100.0)	16	(100.0)	27	(100.0)

from which 8 of the 10 cases emanated. This is the first sign that the much vaunted venture capital market in the Bay area might be producing regionally different results within the total survey sample. Although numbers are relatively small, 8 (30 per cent) of the new Bay area firms founded since 1970 were aided in their birth by venture capital, while this source was insignificant in the British regions. This is a significant marginal regional difference which may aid the birth of small high technology firms in the Bay area. However, it might be argued that the level of venture capital funding in the Bay area is merely caused by the greater number of innovative firms in this area reflecting greater scope for venture capital funding. But this assertion is rendered unlikely by the evidence of Chapter 5, which indicated that the difference in innovation levels between the Bay area and the South East of England is not great. The observed difference in venture capital funding between the two nations is, therefore, more likely to be a reflection of variability in venture capital availability.

9.2.2 *Subsequent employment growth*

It is initially worth noting that small high technology firms are, hopefully, transitory in nature. While there is clear evidence that many small firms cease to operate within the first few years following their birth (Gudgin 1978; Storey 1982; Ganguly 1982) it is also obvious that, by definition, firms that grow cease to be termed 'small'. Indeed, if high technology small firms in the sectors of this study grew rapidly between 1970 and the current survey date, they would not be potential study firms due to the 200-employee ceiling placed on participants. Thus, the following data on employment generated in new firms founded in the five-year period prior to interview are a valid pointer towards the short-term job-generating power of the high technology small firms in the environmentally diverse survey regions. Moreover, the employment growth of such new firms can be taken as a general surrogate for profitability and success in the absence of comparable financial data, which are difficult to compute, both because of confidentiality problems surrounding profits and because of difficulties in financial comparisons between Britain and the United States.

Table 9.2 indicates the number of survey firms founded in each region in the five-year period prior to interview, with the total number of jobs created and the average number of jobs per firm by region. It is surprising that the South East of England recorded a mere 3 new firms during this period. Given that the sample from which these firms were drawn was randomly stratified by size, and was similar in other

Table 9.2 Jobs created in firms less than five years' old by region

(N = 33)	Scotland	South-East England	San Francisco Bay area
Number of firms	16	3	14
Total jobs	251	78	796
Average jobs per firm	16	24	57

respects to its universe and the other regions, it is likely that these new firms are a fair reflection of small firm formation in the South East in the industries and period covered by the study. The other striking feature of Table 9.2 is that firms in the San Francisco Bay area appear generally to grow much faster than their British counterparts. This result bears out observations made during interviews. If the Bay area is compared with Scotland, where the number of new firms is similar, the average size of firm is almost four times as large as the Scottish average figure.

However, the averages of Table 9.2 belie the true explanation for the sharp regional contrasts in employment generation. In fact, the superior performance of the Bay area is largely attributable to three new companies which were founded in the five-year period prior to survey and had grown to employ 125, 150 and 200 workers respectively by the time of interview. The largest growth recorded in Britain for firms in Table 9.2 was 53 employees in a South-Eastern firm. This means that one Bay area new firm generated jobs equivalent to 80 per cent of the total jobs generated by 16 new high technology small firms in Scotland. Although survey numbers are few in Table 9.2, these minority fast-growing firms in the Bay area did not exist in the other regions, and while the number of these firms may be small, the jobs created by them are substantial. Moreover, there is every likelihood that other fast-growing firms in the study sectors in the Bay area had grown beyond the 200-employee ceiling of this study during the five-year study period and thus do not appear in Table 9.2. Indeed, certain recent modifications to the employment figures in the 1982 California manufacturer's directory support this assertion. Firms originally included in the survey on the basis of directory employment figures comfortably below the 200-employee ceiling were subsequently excluded because they had exceeded the 200-employee limit when contacted for interview.

It is generally true that the employment contribution of all small firms to regional economies may be of marginal significance in the short term. However, Table 9.2 hints that in high technology agglomerations such as Silicon Valley, the birth of a small number of fast-growing small firms may have a striking impact on employment. The significance of these minority results at the margin are lent added weight by the known performance of the, now famous, high technology American firms that have grown from the type of new firms indicated in Table 9.2 into large world-famous electronic corporations such as Fairchild, Varian, Mostec and Texas Instruments (Morse 1976). All these firms are less than thirty years old — some much less — but their employment capacity in Silicon Valley alone is measured in tens of thousands. With such impressive employment records it is not necessary to generate hundreds of small firms employing an average of twenty employees each for the next fifty years, but merely two or three Texas Instruments-type firms that subsequently become large. Small firms should not be viewed as an end in themselves, but as large firms in prospect. In Britain there is both a noticeable lack of new fast-growing firms, as shown by Table 9.2, and a dearth of currently large firms of the Texas Instruments type that were small thirty years ago.

9.3 Sources of investment finance

As noted in the case of new firm start-ups, it is no surprise to discover from Table 9.3 that there is an overwhelming tendency in all regions for firms to advance incrementally on the basis of internal resources when funding the main investment needs of the firm. However, Table 9.4 suggests that it is more frequently the highly innovative firm that

Table 9.3 The main source of investment capital by region in the five year period prior to interview

Source of investment capital	Scotland		South-East England		San Francisco Bay area	
(N = 174)	N	(%)	N	(%)	N	(%)
Internal	35	(64.8)	49	(81.7)	45	(75.0)
External	19	(35.2)	11	(18.3)	15	(25.0)
Total	54	(100.0)	60	(100.0)	60	(100.0)

Chi-square = 4.25 $p = 0.120$

Table 9.4 Incidence of product innovation by main source of investment capital

Source of investment capital	Product innovation		No innovation	
(N = 174)	N	(%)	N	(%)
Internal	93	(70.5)	36	(85.7)
External	39	(29.5)	6	(14.3)
Total	132	(100.0)	42	(100.0)

Chi-square = 3.86   *p* = 0.049

seeks external funds, while Table 9.5 shows the tendency for these firms to manufacture high or medium technology output (for a detailed description of the technological complexity variable see Appendix 2). Both these tabulations are statistically significant. Hence, due to the sophistication and innovativeness of the majority of externally funded firms, external sources of investment capital may have a greater effect on the general level of innovation than its incidence in Table 9.3 would suggest. This observation is of particular relevance given that a main aim of this book is to investigate the effect of the local external environment on innovation.

There are two main sources of external capital available to survey firms. The first category, which is available in both Britain and the United States, includes government grants and loans. Clearly, while

Table 9.5 Technological complexity of main product by main source of investment capital

Main source of investment capital	Technological complexity					
	High		Medium		Low	
(N = 174)	N	(%)	N	(%)	N	(%)
Internal	34	(68.0)	34	(65.4)	61	(84.7)
External	16	(32.0)	18	(34.6)	11	(15.3)
Total	50	(100.0)	52	(100.0)	72	(100.0)

Chi-square = 7.27   *p* = 0.026

individual schemes differ between nations, and indeed, between the two British planning regions, government funding is a broadly comparable category. The second category of external finance is private sector funding and, in terms of the current survey firms at least, divides into bank finance and private venture capital.

These various sources of external capital, indicated in Table 9.6, are initially examined together in advance of detailed separate considerations. It was apparent from Table 9.3 that firms in Scotland and the San Francisco Bay area indicated a marginally greater propensity to obtain investment capital externally than firms in the South East of England. However, the breakdown of these external sources in Table 9.6 indicates that they differ in that the main external source in the Bay area was banks *and* venture capital sources, while the dominant source of external capital in Scotland was the local banks alone. The residual 'other' category was mainly comprised of a mixture of various forms of British government assistance.

Table 9.6 Breakdown of main capital investment sources by region

External capital sources	Scotland		South-East England		San Francisco Bay area	
(N = 174)	N	(%)	N	(%)	N	(%)
Profit	35	(64.8)	49	(81.7)	45	(75.0)
Local bank	13	(24.1)	5	(8.3)	6	(10.0)
Venture capital	1	(1.9)	2	(3.3)	9	(15.0)
Other	5	(9.2)	4	(6.7)	0	(0.0)
Total	54	(100.0)	60	(100.0)	60	(100.0)

9.3.1 Government finance

It should be emphasised at the outset that few of the survey firms mentioned government incentives as a major source of investment capital for their firms. However, government incentives can provide a boost to the overall resources of the firm, even when they are not the main source of capital. This is particularly true if such assistance is in the form of grants or low-interest loans. Hence there is, in theory, considerable scope for the use of government incentives in survey firms. In particular, in both Britain and the United States, there are incentives specifically aimed at small firms, while in the Scottish region, the survey firms have access to additional regional development incentives available only in the British development regions.

However, the impact of such schemes will be approached in terms of their *effective delivery* rather than by discussing the plethora of available schemes that theoretically exist to support small firms. It is not surprising to discover from Table 9.7 that the Scottish sub-sample was by far the most extensive recipient of government incentives, with a 78 per cent level of incentive usage in the five-year period prior to interview. The figure for the South East of England was a comparatively low 20 per cent and an insignificant 7 per cent for the San Francisco Bay area. The important role played in Scotland by regional development grants was confirmed by the discovery that 35 firms (65 per cent) of the Scottish sub-sample had obtained regional development grants.

Table 9.7 The incidence of government assistance by region

	Scotland		South-East England		San Francisco Bay area	
(N = 174)	N	*(%)*	N	*(%)*	N	*(%)*
Government aid	42	(77.8)	12	(20.0)	4	(6.7)
No government aid	12	(22.2)	48	(80.0)	56	(93.3)
Total	54	(100.0)	60	(100.0)	60	(100.0)
Chi-square = 72.0	$p = 0.0001$					

In many ways the South-Eastern and Bay area evidence is very similar, as indicated by the low use of government incentives in Table 9.7. The reasons for not utilising schemes in these regions were many, but the most frequently quoted general reason was that various forms of government red tape had either inhibited or caused the abandonment of attempts to obtain assistance. On the subject of red tape, there were considerable similarities between the South-Eastern and Bay area firms since most of the schemes available in these areas, unlike Scotland, were discretionary, and hence subject to a more extensive decision-making process. In the South East the MAPCON scheme, designed to provide grants towards consultancy on the feasibility of microprocessor application in a firm's products, had caused problems of access where, of the 21 firms attempting to obtain the incentive, only 5 had succeeded. In the Bay area the pattern was much the same. In Table 9.7, the 4 Bay area survey firms obtaining incentives had all made use of various Small Business Administration (SBA) loans. However, a further 22 firms in the Bay area stated that they had sought, but failed to obtain, SBA loans. In instances where discretionary incentives were involved in both regions, the most common complaints concerned the bureaucracy involved in the application procedure and, subsequently, the long delay in making a decision on a loan.

9.3.2 Bank finance

On the surface there are substantial differences between the British and American banking systems. In particular, the British system is based on four main banks, while the United States, due to a legacy of restrictive legislation, has very few national banks and is instead served by many thousands of locally based banks. However, in practice, the international nature of banking practices in terms of both interest rates and loan conditions means that the effects of both British and American banks at plant level in the small firms of this study are very similar. For example, it is likely that the policies of the British and American governments in recent years have helped create high interest rates which have impeded the progress of small firms in both countries.

The plant-level experiences of the survey firms have been indicated in Table 9.6 which, apart from the Scottish instance with a 24 per cent level of usage, confirms that the effects of banks as a main source of investment capital in the survey firms was negligible in the South East of England (8 per cent) and the San Francisco Bay area (10 per cent). The relatively high 24 per cent level of bank funding in Scotland may well reflect a genuine attempt by indigenous Scottish banks to play a more vigorous role in industrial financing than their nationally based counterparts. This was certainly the view of a recent report of the Monopolies Commission (1982) on the Royal Bank of Scotland. However, the overall 14 per cent level of bank funding in the survey firms is low given the generally high technology nature of the firms in the total sample.

Conversations with individual executives in both Britain and America yielded a surprisingly similar range of comments on the value of bank funding. Indeed, many of the previous comments on the problems of government incentive delivery also apply to banks. It is often felt that the processors of loan applications are financially, but not technically, qualified personnel who do not see the sales potential of proposed innovations for which finance is required. In common with comments on government funding, criticism centres on the length of time taken to decide on a loan application. In this context, responses to a question on loan application refusals (Table 9.8) are enlightening. There was a very low level of 26 (15 per cent) of refusals in the total sample, with no regionally significant variations. However,

Table 9.8 Bank loan application refusals by region

Bank loan application (N = 174)	Scotland		South-East England		San Francisco Bay area	
	N	(%)	N	(%)	N	(%)
Refused	8	(14.8)	11	(18.3)	7	(11.7)
Not refused*	46	(85.2)	49	(81.7)	53	(88.3)
Total	54	(100.0)	60	(100.0)	60	(100.0)

* This does not imply that these firms applied for or obtained bank loans.

this meagre aggregate statistic is worthy of further contextual comment. Conversations with many executives throw light on the apparent paradox in which Table 9.6 indicates a generally poor level of bank finance usage while Table 9.8 confirms that there is also a very low level of loan application refusal. The relationship between a small business owner and his bank manager is in many cases reasonably close. In the years since formation, the owner develops a working relationship with the manager that permits occasional overdraft facilities which are easily agreed. However, the closeness of the relationship also means that small firm owners develop a good working knowledge of the terms and conditions under which a bank manager will loan money.

Hence innovative firms in need of investment finance often do not approach banks because they feel, rightly or wrongly, that their application will be refused, with some loss of face. The arguments of Chapter 7 on research and development cycles are relevant here in that the most innovative and fast-growing firms often undergo periods of loss making during the performance of expensive research and development in advance of a product launch. However, banks frequently appear more concerned with a healthy current bank balance than with the future innovation and subsequent growth potential of small firms. At a practical level, beyond their public protestations that investment in high technology firms is a fundamental objective, banks are often wary of innovative high technology firms because they find the technical potential of a proposed product development difficult to judge, while all the financial evidence available to accountancy-oriented evaluators reveals a poor financial position brought on by the cost of research and development and its detrimental effect on short- or medium-term profits. In such circumstances, firms with confidence in the ultimate financial success of their technically prestigious product developments bypass banks and seek other sources of investment funding. Conversations with executives in both Britain and the United States confirm that banks have a poor reputation for the extension of investment capital, and these opinions both bear out and help to explain the low general level of bank funding together with a low level of loan application refusal.

9.3.3 *Venture capital*

Venture capital and its effect on the survey firms is worthy of inclusion here, not only to confirm its significant marginal effect on a minority of businesses, but also because it is necessary to put in context many of the myths surrounding this form of investment finance. In terms of both start-up capital (Table 9.1) and main investment capital (Table 9.6), it is clear that venture capital is not a major source of investment capital in any of the survey regions. However, it is also abundantly clear in both tables that the pre-eminent origin of venture capital is the San Francisco Bay area of California. It may be recalled that 30 per cent of the new firms founded in the Bay area since 1970 were established with venture capital as a main source of start-up funding, while 15 per cent of the Bay area total sample acknowledged that venture capital was a main source of investment finance during the previous five years. These marginal statistics compare with a virtual absence of any private venture capital sourcing in the British regions. Thus it is clear that the impact of venture capital is limited to high technology small firms in the San Francisco Bay area and is not evident on the scale that many media commentators would suggest. This view is confirmed by other recent research (Bullock 1983). However, it is also clear that venture capital does perform a useful function at the margin by making another source of investment capital available to entrepreneurs, particularly those with good prospective projects who are in need of a rapid injection of capital.

Moreover, there is much anecdotal evidence among Silicon Valley entrepreneurs that venture capitalists are particularly good at 'picking winners' and display a willingness to back their judgement rapidly with large amounts of cash. Because many of the venture capitalists are ex-businessmen with technical training and a practical knowledge of production, they are eminently qualified to judge both the business acumen of potential recipients of venture capital and the technical viability of the product development for which it is sought. Since the venture capitalists are the individuals who hold the purse-strings, they are able to advance money very quickly once they have made a decision. In many ways, the interest of a well-known venture capitalist or of a consortium of venture capitalists is, in itself, a guarantee of success in that once such prestigious investors become involved in a small high technology firm, they have a vested interest in ensuring its success, since their reputation depends on the transformation of promising small firms into large 'winners'. Reputation is important because many venture capitalists may act as 'front men' for large flows of institutional capital from banks and pension funds, where their expertise is essential in order to build up a portfolio of local small Bay area firms. Each institution might contribute, say, one million dollars. Since the venture capitalists are selling their expertise, it is bad for business if a firm they have backed goes to the wall. Indeed, at the time of interview one survey firm in the Bay area was being 'propped up' by a venture capitalist consortium for this very reason. In general terms, however, this example is atypical, and the short-term profits from the aggregate growth of a portfolio of fast-growing high technology small firms can be very attractive.

None the less, certain drawbacks are also associated with venture capital. The most serious problem resulting from the involvement of venture capitalists is the fear that internal control may be lost, at least in part, to the venture capitalist. This fear is manifested when equity is exchanged for investment capital, as is usual in such circumstances. Clearly, there is a danger that subsequent growth, spurred on by the growth prompted by the injection of cash from previous equity sales, may tend to lead to further capital shortage, which in turn leads to further equity transfer to the venture capitalist in return for more cash as the product life cycles, investment cycles and firm size grow as indicated in Figure 7.1. A point may easily be reached where this trend is irreversible and the original small firm owner is forced to sell his remaining equity in the company, albeit in return for a healthy bank

balance. Indeed, many venture capitalists are themselves entrepreneurs who have 'sold out' or been 'bought out' in the past, thus
accruing the capital and expertise necessary to begin their own venture
capital businesses.

Regardless of the size of the equity holding, the main objective of
venture capitalists is to purchase relatively cheap equity in rapidly
growing corporations in order that the shares may later be sold at a profit
when the firm is sold privately, or publicly quoted on the stock exchange.
Clearly, a controlling interest is a desirable ultimate objective for the
venture capitalist since it ensures direct management control over his
investment, irrespective of whether the 'management' remains nominally
led by the firm's founder. However, the small firm executive is aware of
the objectives of venture capitalists and, even in Silicon Valley, many
owners expressed a desire to move forward at a slower pace than
externally funded firms on the basis of their internal profits, thus avoiding
the sale of equity and surrender of the total control of the business.

However, at an *aggregate* regional level, the efforts of local venture
capitalists are beneficial to high technology agglomerations in two
major ways. First, although overall use of venture capital may not be
extensive, venture capitalists broaden the potential range of external
capital investment sources for industrialists looking for investment
finance. And, notwithstanding their motivations, the substantial
amounts of hard cash that venture capitalists make available at short
notice can be very beneficial to firms with growth potential. Moreover,
while the potential loss of ownership may be a legitimate fear of small
firm owners, the injection of large amounts of venture capital into the
fast-growing small firms of a region will produce employment
increases, regardless of who ultimately owns the business. Second, it is
likely that the eventual venture capitalist owner will be local, since
venture capitalists in the Bay area are predominantly local men
who are ploughing back profits earned in the local area into new
businesses in the same local agglomeration. Since their main expertise
lies in knowing what is being developed within the high technology
local agglomeration, of which they have specialist knowledge, there is
every likelihood that they will restrict their investments, and hence
level of risk, to the known local business environment. These venture
capitalists may be seen as an integral part of the high technology
business agglomeration, since they recycle the profits from previous
local enterprise. Morevoer, as already mentioned, venture capitalist
organisations also attracts external funds from national financial
institutions who want to 'get in on the act'.

The behaviour of Bay area venture capitalists is important, not merely because of their direct contribution to the funding of high technology small firms, but because of the lessons that may be learned from them by the more formalised public and private providers of investment capital to industry in other less prosperous regions.

9.4 Innovation finance and small firm production niches

It is clear that any substantial programme of innovation in small firms demands a considerable commitment of investment capital to fund research and development costs. In many cases, programmes of innovation in high technology sectors lead small firms to seek financial support from external individuals or institutions. However, in these concluding passages it is helpful to set the warranted concern for fast-growing high technology small firms in a broader policy context, since the policy instruments for the encouragement of *both* fast- and slow-growing small firms are predominantly financial.

This approach begins with a consideration of the business types that comprise the high technology small firm sample used in this study. It is initially useful to re-emphasise that not all 'high technology' firms, defined as such because they fall within a high technology MLH or SIC categorisation, are necessarily high technology or *innovative* businesses. It is important to consider these differences, since they affect the applicability of innovation finance to individual small firms. While, in reality, each individual firm is unique, at a generalised level three major categories of firm are now proposed in an attempt to explain the ways in which the impact of innovation finance has varying small firm applicability.

The first type of small firm identifiable within high technology sectors is the subcontract firm discussed previously in Chapter 3. This firm is not product based but depends on customers for the main technical specifications of its product. Research and development is low or non-existent in such a firm and, when it occurs, is mainly concerned with the application of new process techniques. Indeed, a great advantage of this type of production is the low research and development and marketing costs that result from the absence of a product. These firms formed a major part of the group of survey firms that, according to the evidence of Chapter 5, did not introduce product innovations. Clearly, in such circumstances, the relevance of innovation finance is minimal.

The second category of firm is one that is very common in any economy, but which rarely receives much comment or interest. This firm is a 'long product life cycle' based firm. Such a classification does not contradict the high technology status of the general high technology category from which this type of firm emanates. Many long life-cycle products are classified within high technology sectors because they are an integral part of high technology production (for example specialist thermometers, pressure gauges and heat exchangers). While these products are ageing, and while demand may be falling slowly residual markets ensure a good level of business for such small-scale producers. Indeed, in a sense, this declining market protects such firms from the competition of larger producers. Typically, after the initial burst of innovation which developed the firm's product, research and development declines to a low level. This type of business is frequently characterised by a father–son type of evolution, where the father developed the product many years ago and the son has taken on the business, which remains viable. However, product development stagnates because the son does not possess the technical acumen of his father.

The relevance of innovation finance to this type of firm is again marginal. Most development work, if it occurs, relates to the refurbishing of the existing product. Management tends to be defensive, and attempts to minimise the risk by avoiding radical technical change. Typically, the objective of such executives tends to be no more than a good living wage. In many cases, such firms do not have a new product idea waiting for development cash. Many of these firms would also fall into the category of survey firms which, as described in Chapter 5, did not introduce new product innovations or they would fall, marginally, into the category of innovating firms which revamped their traditional product. The injection of innovation-oriented investment finance into such firms, therefore, would not be aimed predominantly at the development of a new product idea *per se*. More fundamentally, investment would need to be directed at a revitalisation of the basic approach of the firm towards innovation through a long-term attempt to develop an internal research and development capacity from which new products might evolve. Clearly, attempts to urge such firms to become more innovative through a radical revision of their fundamental approach to business is a particularly risky procedure, partly because of the danger of over-extending the capabilities of the firm's management, and partly because there is frequently no embryonic product on which to focus a decision on capital investment. Hence, using capital investment to stimulate innovation in such firms is a particularly nebulous business.

The third category of firm is the archetype that is most commonly associated with high technology production. This is the firm, frequently described in previous chapters, with a high technology product base that involves rapid and short product life cycles. This firm is typically newer with aggressive management which often makes use of external capital sources in the ways outlined in this chapter. As discussed in Chapter 7 in a research and development context, such firms, due to the frequency of their product life cycles, need to perform a large amount of research and development which frequently means, due to offset 'investment' and 'research and development' cycles, that the firm is periodically put under considerable financial stress (see Figure 7.1). Clearly, this is the type of firm most in need of innovation finance. Moreover, since the approach of the firm is ambitious and based on evolving products, grant and loan applications to external loan agencies may be based on a tangible technical concept. Such characteristics imply that high technology product-based firms have great growth potential.

An overview of these typologies suggests that within high technology industry, there are a wide range of firms with diverse financial needs. Therefore, it is essential that any government policy based on financial aid towards innovation in small firms should take account of the varying aptitude of individual firms. In keeping with the British comprehensive school ethos, where all levels of ability are simultaneously encouraged, each individual firm should be encouraged to develop whatever potential exists. In truly high technology, aggressive, product-based firms, the rate at which capital can be gainfully employed through extensive research and development will be rapid, while small subcontract firms might be adequately assisted by a grant with which to purchase, for example, a second-hand drilling machine (grants for second-hand equipment are currently not available in British assisted areas). Thus, policy design and delivery should be flexible, with assistance packages that cater for the smallest needs while remaining able to accommodate *adequately* the resource requirements of the fastest-growing high technology small firms. For it is worth re-emphasising that both the evidence of this chapter and the known growth of previously small high technology American firms such as Texas Instruments indicate the employment benefits that might accrue to regional or national governments from the vigorous assistance of high technology small firms with genuine growth potential. Much of the resistance in survey firms to current government assistance, as described in Section 9.3.1, would be largely removed if future incentives were delivered in a more flexible manner with a delivery approach credible to small firm owners. Indeed, the delivery system might adopt many of the strengths of the venture capitalist's approach. However, a further detailed consideration of high technology small firm funding and the relative roles of government incentives, bank funding and venture capital is reserved for the following concluding chapter on theory and policy.

10 Conclusions

10.1 Conclusions for theory

10.1.1 High technology industry and the principle of agglomeration

The principal objective of this book has been to examine a diversity of regional industrial environments to establish the effects of variations in local external resources on internal small firm innovation. In presenting these concluding comments, however, it is useful initially to emphasise the links between local resource advantages and the concept of agglomeration. In Chapter 4 and at various points during the analytical chapters, there has been acknowledgement that the term 'local resource advantage' is virtually synonymous with the concept of agglomeration economies. Taken at their broadest level of definition, agglomeration economies include the local advantages of rich labour, linkage, technical information and financial resources. Thus, if the study concludes that local resource advantage exists for small high technology firms in the most innovative San Francisco Bay area, it is implicit that this area is also one of agglomeration advantage. Such a conclusion would have a regenerative effect on the theory of agglomeration, which had seemed, certainly during the 'footloose industry' dominated 1960s, to have become an inapproriate means of conceptualising the location requirements of 'modern' industry. Indeed, due to the future growth potential of the industries concerned in this study, these assertions would reinstate the principle of agglomeration as a key issue in the evolving industrial location debate.

On balance, this book must conclude that there is evidence of agglomeration economies in the most innovative regions of the study. With the exception of local technical information contacts, data on the remaining resource inputs to the innovation process show either signs of agglomeration advantage in the two more prosperous regions of the South East of England and the Bay area together, or in the Bay area alone.

Significantly, the evidence on input linkages provided strong evidence on agglomeration advantage, both because of the preponderance of local linkage servicing in the highly innovative Bay area, and because of negative impact of a poor input linkage infrastructure on innovation in the Scottish development region. While high technology industries differ from the vertically disintegrated industries in the agglomerations of the past by *not* being dependent on local customers to any great degree, their strong dependence on local suppliers in the densely packed Bay area industrial complex is an archetypal sign of agglomerative advantage. The locationally constraining technical dependence of these firms on local specialist suppliers must be a direct facilitative spur to innovation and subsequent growth.

If a broad definition of local agglomeration economies is maintained, investment capital would clearly qualify as a particularly rich local source of agglomeration advantage. In particular, since such investment capital fuels the innovation process, variations in its local availability will clearly directly affect regional innovation levels. Hence, it is no surprise to discover a significant marginal impact of venture capital in the San Francisco Bay area of California, particularly in the fast-growing firms. This marginal Bay area advantage was sharpened by a virtual absence of this type of local financial resource in the British regions.

Ironically, the evidence on *higher* labour shortages in the two more prosperous South-Eastern and Bay area regions when compared with Scotland, only further supports the concept of agglomeration. Modern experience suggests that today's agglomerations are areas with intensely competitive labour markets where 'overheating' of the local economy forces wages upwards. The phenomenon of a large, local, highly skilled labour supply, but higher labour demand, produces a situation of both dense skill concentrations of blue- and white-collar workers *and* acute labour shortages. This phenomenon supports the agglomeration thesis for such areas, since the continued existence of firms in these locations is clear proof of their willingness to pay inflated wages for essential workers, in order to keep other agglomeration economies available locally (Nicholls 1969; Alonso 1971). Put simply, labour shortages and high wages in the local area are a bearable price of the success that results from location in the agglomeration. To some extent, labour shortages and high wages in a local economy are logical phenomena since, if it is assumed that agglomerations possess competitive advantage, it should be no surprise to find them expanding, thus producing labour demand, labour shortages and consequently high wage levels.

One further general observation on the impact of agglomeration economies on high technology industries is important, particularly with regard to the Bay area of California. The specific contribution to agglomeration advantage from individual resource factors may be of varying locational significance when taken alone. However, much of the strength of broadly defined agglomeration economies derives from the cumulative effect of the multiple advantages of a single location. There is a sense in which the whole of the agglomeration advantage is greater than the sum of its parts. A critical mass of agglomerated industry is created that triggers circular cumulative advantages which increase over time. These growing advantages in part stem from quantifiable economies in production, but also result from a belief in the minds of firm executives that they are 'at the centre of things'. It is certainly true that many executives in the Bay area believed their local economy to be the best in the world for high technology small firm production. Clearly, this may be true in specific instances. However, whatever the objective facts, the belief that the Bay area of California is an optimal location for existing and new high technology production is a significant contextual point for the following consideration of policy. These beliefs are a major force towards an industrial inertia which benefits such areas at the expense of depressed regions with aspirations of acquiring high technology industrial production. As the earlier growth of iron and steel and textile production in specific locations has indicated, the effects of agglomeration advantage can be a major barrier to an equitable national distribution of production.

However, mention of past industrial agglomeration raises a pertinent element of doubt over the prospects for future high technology industrial growth, not merely in those areas seeking to acquire it, but in the very heart of existing complexes such as Silicon Valley. For example, it might be argued that the present 'boom' areas of high technology industrial expansion will be the declining industrial problem areas of tomorrow. Indeed, at the height of production in the Lancashire or New England textile areas there was little anticipation of their recent dramatic decline. A consideration of the long-term sustainability of high technology industrial agglomerations may be handled to good effect by resort to the product life cycle concept, applied at industry level.

10.1.2 *Macro and micro product life cycles*

Perhaps the most attractive aspect of the original product life cycle theory is its temporal dimension (Vernon (1966). While much theory pertaining to industrial location and innovation relates to a static event, the product life cycle, applied at the micro level of the firm, allows for the fluctuating fortunes of a business as it grapples with the problems ensuing from interaction with a largely hostile business environment. The product life cycle of the individual firm has been the basis for discussion at many junctures during the empirical passages of this book. It has helped to explain the behaviour of firms as they attempt, in the face of limited funds, to match the research and development and labour commitments of the firm to the needs of the marketplace. It is a useful contribution to the current discussion of agglomeration, however, to view product life cycles at a macro sectoral level in which the individual products of specific high technology electronics firms can be aggregated into a general product category for the industry as a whole, in much the same way as the woollen textile industry is an aggregation of individual products such as worsteds or tweeds. None the less, at this aggregate level the product life cycle of an industry remains similar to that of the single firm, with phases of growth, maturity and decline.

The acknowledgement of the propulsive effects on sectoral growth from technological innovation is central to an understanding of the works of such great theoretical economists as Kondratiev (1935) and Pirroux (1955). However, it is central to the current argument to note that these industrial sectors have often been physically concentrated to reap agglomeration economies in regions that have frequently suffered when the upward movement and crest of their industry-level product life cycle was followed by maturity and subsequent decline. England abounds with industrial areas that were founded on the undeniably propulsive effect of Kondratiev's steam-driven innovation wave of the nineteenth century, only to face acute problems of industrial decline in the mid-twentieth century. Hence, the historical evidence on the long-term survival of industrial agglomerations does not augur well for the new high technology industrial concentrations.

However, perhaps the most encouraging evidence to indicate that new high technology industrial agglomerations are less likely to fail is their great capacity to create new products and industries with which to ensure continued survival. The common feature of all the high technology industries is their uniformly high commitment to research and development, which is a good broad definition of high technology industry. This intense research and development commitment, both at the level of the individual high technology firm, and at an aggregate level when they are agglomerated, ensures rapid technological change and the acquisition of new leading-edge technology, either through indigenous development in the agglomeration, or through the attraction into the agglomeration of externally discovered ideas. The high quality of the development and production workers in high technology agglomerations is an attractive force to any external entrepreneur seeking a location in which to develop a high technology product.

The evolution of industries based on new techniques, from silicon-based semi-conductors to bio-engineering technologies, seems to ensure the growth of high technology agglomerations in the foreseeable future due to the regenerative effect of new industry-level product life cycles on the agglomeration. Hence, a great strength of the new high technology industrial agglomeration is that it is based not on agglomeration economies resulting from the manufacture of a single industry, such as cotton textiles or steel, but on the output of a highly skilled research and development and production-oriented workforce that can adapt to totally new technical innovations and production concepts. Thus, it is unlikely that these new agglomerations will suffer the problems of innovation, stagnation and subsequent decline common in their historical predecessors.

10.1.3 *High technology industry and the growth pole concept*

There is now substantial evidence to suggest that high technology industrial agglomerations possess the ability to regenerate industrial activity through multifaceted product evolution on a broad technological front (Cooper 1970; Little 1977; Bullock 1983). This potential for current and projected growth has prompted planners concerned with the promotion of manufacturing employment growth in declining industrial regions to seek a core of high technology industry as a base for similar subsequent industrial expansion, both within the initial firms, and from spin-off enterprises that might logically be expected to evolve from these firms as they achieve large size and a high level of research and development and shopfloor workers. However, while modern planners might view this approach as a new departure, it is little more than a resurrection of the growth pole principle much in evidence in the 1960s.

The initial growth pole principle was postulated by Pirroux (1955) in a mainly aspatial context in which he argued that propulsive sectors of an economy might act as foci for national industrial growth, although he did not develop the geographical implications of his new approach (Nicholls 1969). Later writers expanded the concept by providing a clear geographical, and thus a physcial planning, basis (Darwent 1969; Thomas 1975). While a whole range of variations on this growth pole theme was considered or put into operation in various national contexts, perhaps the nearest spatial adaptation of Pirroux's original concept was the approach that argued for industrial growth based on a propulsive industry in a specific physical location. It was argued that certain industries, for example iron and steel production or oil refining, might act as providers of basic materials for a wide variety of downstream engineering and petrochemical industries. Such generic industries would add value to these basic outputs in the vicinity of the growth pole plant. However, as Chapman (1973) pointed out in the case of oil refining, the concept of linkage between plants, based merely on the premiss that they were proximate, was naive in that it ignored many technical and organisa-tional barriers to the free flow of goods between spatially concen-trated, but corporately separate, firms that were often competing in similar sub-sectors of the same industry.

A central problem of the spatial adaption of the growth pole principle was therefore the inability of the propounders of this approach to suggest convincingly an industry that would act as a viable pole around which subsequent growth might take place. However, the expansion of the Silicon Valley and Route 128 complexes in the 1970s has provided clear proof that high technology electronics-based industry can provide a successful pole for rapid and sustained industrial growth in a sharply defined polar (or linear) physical context. As already mentioned, it is doubtful whether planners in the United States or Europe were conscious of resur-recting the growth pole theory in their attempts to replicate such spontaneous growth, but it is evident that the recent clamour among development agencies from Cleveland, Ohio to Liverpool in England to promote the birth and growth of their own Silicon Valley, gulch, glen, wadi or science park is no more than a rebirth of the growth pole planning principle based on a high technology propulsive industrial node. None the less, the objective of creating high technology industrial growth poles with which to effect industrial structural change and employment growth is a laudable goal whether it is in older declining industrial areas through, for example, the Scottish

Development Agency in Scotland, or in agricultural areas, as evidenced by the Irish Development Agency in Ireland.

Moreover, there are specific attributes of high technology production that particularly recommend it as a suitable node for industrial expansion in depressed regions. First, the preceding paragraphs suggest that such industry is likely to prosper and have a good medium- to long-term life expectancy. Further, it is critical to the success of this growth pole that such a concentration *is* likely to act as a source of spin-off firms which will further contribute to the diversity and depth of expertise available in the expanding high technology industrial node. Second, if a core of high technology production can be established in older declining industrial areas, the marginal agglomeration advantages of linkages and labour skills such a node of high technology production generates locally when compared with regions without such a concentration, will be an attraction for any mobile production relocated from established agglomerations, particularly if government incentives are available. However, as intimated in the previous discussion of agglomeration advantage, development agencies will not readily entice high technology industrial production away from the economies of existing agglomerations. None the less, such concentrations clearly generate disadvantages of high rents, local taxes and labour costs that may render lower-cost locations in declining regions more attractive, particularly when the government incentives available in such areas are taken into account. For there will always be a small amount of mobile high technology production resulting from industrial expansion in both intra- and international contexts.

10.2   Conclusions for policy

10.2.1   *Siting a high technology industrial growth pole*

It has been argued that the high technology industries of the type common in Silicon Valley might act as a suitable focus for an industrial growth pole in depressed regions. The growth pole concept is particularly pertinent to these high technology forms of production because a feature of their 'spontaneous' growth has been the physical intensity of the subsequent industrial concentrations. For example, the Silicon Valley industrial complex fits easily into a rectangle measuring twenty by ten kilometres. Thus, development agencies seeking to replicate the success of Silicon Valley might concentrate assistance into narrowly defined spatial zones, a concept that lay at the heart of much growth pole planning of the 1960s. It would seem reasonable to designate sites for high technology industrial agglomerations at environmentally attractive locations with the best facilities for growth, possibly on the outskirts of a major conurbation near to an airport and a main road link. Ideally, the chosen site would contain little previous industrial activity. Using the British spatial scale as a yardstick, the size of the chosen pole could be well below the sub-regional level, perhaps on an area of land set aside for a predetermined mixture of new industrial and residential expansion of no more than thirty square miles, to be taken up as and when demand dictated. The industrial part of the development might be allowed to expand from a central core of various-sized plots, with a mixture of 'off the peg' units and vacant sites for purpose-built factories of various sizes.

In this context, currently similar schemes would not satisfy the ethos of this proposed development, on two counts. First, while the existing enterprise zone policy possesses many development features similar to the proposed growth pole approach, it suffers from the predominant disadvantage of operating in areas of existing industrial dereliction and decline. Clearly, attempts to attract inward invest-ment and promote indigenous industrial growth would be greatly impeded by this initial environmental disadvantage. Second, the science park concept, which is currently very popular in many Western countries, suffers from the formalising influence of the estate approach to industry. Indeed, many of the 'science parks' are merely old industrial estates with a new signboard at the gate. The park or estate approach has many drawbacks, not least of which is the unwillingness of many large corporations to be considered within the provisions of a preconceived development in general, or factory units in particular.

With appropriate planning permission, the proposed high tech-nology industrial growth pole could be allowed to expand at its own pace within the flexible confines of an overall structure plan for the area. However, it would be important that this zone be restricted to *truly* high technology firms, both because local high technology industrial agglomeration economies will only accrue if the incoming firms are based in high technology forms of production, and because the whole concept of this flexible growth pole, or mixed industrial and residential housing, depends for success on the *environmental acceptability* of the forms of production introduced. It is clear from the speculative development in Silicon Valley that industry and private housing can sit well together in a loosely structured, but integrated, manner. A clear advantage of the introduction of high technology industry into a previously non-industrialised area would be its ability to provide local prosperity without destroying the quality of the local environment. Moreover, the integration of such industry into a local area with an attractive environment would enhance the prospects of attracting the executives of large corporations and the skilled white- and blue-collar workers that would staff the ensuing factories.

It is essential that this proposed high technology growth pole should not be a vehicle for political expediency. The evidence of this book suggests there are intense and increasing agglomeration forces of industrial inertia in areas of existing high technology industrial production. These constraints will not be broken by the dubious attractions of a science park set up in a disused steel plant in an area of industrial dereliction. It may be politically expedient to appear to be contributing to industrial regeneration at the point of maximum need following the closure of a steel plant or car factory. However, the medium-term benefit of such an action will be minimal, since the project is likely to fail. All depressed regions have within them areas with high-quality environments, and one of these should be the chosen area for a high technology industrial growth pole. Planning for these zones should be flexible and facilitative, with aggressive marketing of the chosen area. Local and national government agencies must not ignore the specific needs of high technology industry if they wish to be successful in its attraction.

10.2.2 Launching the growth pole

The largest single problem facing the establishment of a high technology industrial growth pole is that of credibility. Clearly, some form of 'critical mass' of high technology production should be established as rapidly as possible with which to attract further inward investment and from which indigenous firms might 'spin off'. If the full range of government incentives were focused within a depressed region on the environmentally attractive pole discussed in the preceding section, adjacent to blackspots of severe industrial decline, there would be a real chance of attracting some new or existing high technology industrial production from the 'over-heating' high technology industrial agglomerations. The attractive-ness of such an area would be enhanced by the previously mentioned exploitable negative aspects of high wages, and high costs for both factory and private house building, in existing agglomerations.

The same high levels of congestion in prosperous areas, combined with the anticipated attractiveness of the physical environment in the proposed growth zone, might promote the success of governments' efforts towards the relocation of institutional and corporate research and development activities to the new growth zone. Indeed, in Britain, Department of Industry's current Offices and Service Industries Scheme (OSIS), partly aimed at the relocation of corporate research and development activity away from the South East of England, might meet with more success if an environmentally attractive high technology industrial complex within depressed regions could be recommended to potential movers. One of the major inhibitions to the relocation of high technology industrial production and research and development activity is the fear among individual movers that they are in danger of isolation, both in physical and in job-mobility terms, if they personally relocate. The prospect of a growing high technology industrial complex in which 'job hopping' might be possible would reduce such fears. However, it is acknowledged that in the early years of the zone's growth, vigorous efforts would be necessary to break the vicious circle in which an absence of a quorum of high technology skills prevents their acquisition.

Apart from the inward movement of production, it might be expected that an environmentally attractive growth zone with full government regional incentives would attract a number of existing small and larger high technology firms from within the region. Indeed, the zone might be based on an existing, suitably sited large manufacturing firm. However, it is clear that a build-up of research and development-oriented high technology manufacturing units would be necessary before high technology agglomeration economies began to assist further growth. Similarly, second-generation spin-off firms would not contribute to indigenous growth in the early years. Hence, it is clear that a medium- to long-term planning approach should be adopted, in which significant results should not be expected before at least a ten-year gestation period. The key agglomeration advantages of a highly skilled white- and blue-collar workforce and a wide range of input linkages in both materials and services are sadly lacking in depressed industrial areas. The changing of industrial and subsequent labour market structures through high technology industrial expansion is a slow and laborious process, but worthwhile due to its future growth potential. Facilitative growth pole planning strategies, directed at industrial structural change at the sub-regional level, will help create a conducive environment for high technology innovation and growth in incoming and indigenous firms. Further policy proposals for the encouragement of innovation within individual small high technology firms are considered in the following pages.

10.2.3 *Policies for the encouragement of innovation in small high technology firms*

There is considerable scope for policy at the specific level of the high technology small firm considered in this book. The central importance of research and development to the innovation process in individual firms, and its subsequent effect on innovation, has been a recurrent theme in all the preceding, empirically based chapters. Indeed, the locational insignificance of external technical information acquisition was largely attributed to the overwhelming importance of internal research and development and its close interaction with production on the shopfloor. The importance of internal research and development is also reflected in the commitment of substantial amounts of internal or borrowed capital with which to fund this effort. Thus, since research and development is clearly central to any explanation of innovation and growth in small high technology firms, and further, since such a process is mainly based on the injection of investment finance, there is obvious scope for the impact of research and development-oriented government funding policies.

The evidence of Chapters 2 and 5 has indicated that the depressed regions of Britain do not possess their proportionate share of high technology sectors, and that those firms from innovative sectors that do exist in such depressed regions are proportionately less innovative than matched firms from the South East of England. None the less, it remains true that any depressed region, whether it be the North East of England or a depressed part of the manufacturing belt of the United States, will have a number of high technology firms with growth potential. But these depressed regions, by definition, rarely create the internal wealth from industrial growth that facilitates the emergence of local venture capitalists with industrial experience who are willing to fund such firms. However, government aid available in such depressed areas might be directed at these firms, with the specific remit of filling the venture capital gap by providing investment capital directly aimed at research and development which, in turn, should greatly promote the production of new products, expanded markets and employment growth. The limited evidence from this book has indicated that venture capital provides a useful service in the San Francisco Bay area of California through the rapid provision of investment capital for fast-growing firms. In this context, it is of further significance that survey evidence from both Britain and the United States has indicated that banks are not particularly forthright in the investment capital funding of small high technology firms. While the gap left by institutional private sector funding in the Bay area is filled, at least in part, by the activities of venture capitalists in various guises, the gap remains in other parts of the United States and in the whole of Britain. In this book, statistics and anecdotal evidence from both Britain and the United States indicates that it is mainly the fastest growing firms with the greatest potential that are most frustrated by an inability to raise venture capital. For example, the most successful British obtainer of small firm investment capital in the survey had obtained all the major relevant grants and loans including National Research and Development Corporation (NRDC), Industrial and Commercial Finance Corporation (ICFC) and Microprocessors Application Scheme (MAP) assistance. However, this managing director vigorously asserted that, given *adequate* funding his business might have been three times larger.

Venture capital is centrally important to this study not only because of its significant marginal importance to innovation and growth in Bay Area firms, but also because the behaviour of venture capitalists in funding small high technology firms points to many lessons that might be learned by institutionalised funders in both the public and private sectors. First, the rapidity with which capital is advanced (or refused) by venture capitalists is enlightening. Because of the short product life cycles common in high technology industry, it is essential that decisions on loans be made in a 'next week or not at all' spirit. Indeed, it is important that a firm not ultimately obtaining funds be told of a decision quickly. Venture capitalists compress the decision process to a point where, in many cases, the decision to fund is made virtually on the spot, and funding is extended within days rather than months. Clearly, it is important that the evaluators of proposed investment funding be able to extend the finance. Here the element of risk is clearly higher than for longer term evaluations. However, venture capitalists reduce the risk of individual firm failures by building up a portfolio of small firms in which various levels of equity stakes may be taken. Thus, the failure of individual firms is more than compensated for by other successes and *aggregate* profits are substantial. Risks are further reduced by the personal experience of the venture capitalist, who is frequently an ex-manufacturer with considerable business and technical acumen.

Much of this behaviour could be replicated by public and private institutions in their efforts to provide investment capital for research and development in small high technology firms. Government agencies in depressed regions could make considerable advances towards ensuring that the minority of high technology small firms that are born in largely hostile innovation environments are compensated by a locally superior environment for investment finance, with regard both to terms offered and speed of service. Importantly, the level of bureaucracy inherent in such an objective should be kept to a minimum, and the level of risk should be *deliberately* increased to ensure that worthy firms are not starved of funds for expansion in order to maintain a spuriously high success rate for loan approvals. As in the venture capitalist case, risks could be reduced by ensuring that the evaluators possess technical knowledge, and, through the establishment of a portfolio of firms, by taking the equity stake as return for the capital.

To this end, an agency might be established in a depressed area with the sole remit of providing venture capital for high technology small firms. The agency could be established with a lump sum of capital from national or local government, or from combined private and public sources, and would be empowered to invest the money in suitable firms, taking an equity stake or an equity stake option in return for the funds. The agency might also develop technical and business 'back-up' services which would assist the evaluation of initial investment in suitable firms, and aid their subsequent growth. This organisation would be mainly self-financing after the initial injection of capital, since profits from the sale of equity stakes in successful firms would be ploughed back into the central fund. Such an approach would not be politically contentious, since equity stakes in small firms that became established might be sold back to the private sector in order to provide funds for further assistance. Apart from the self-financing nature of this approach to small firm assistance, the direct government revenue from tax paid by the growing firm and its employees, and the reductions in unemployment benefit resulting from the expansion of such small firms in depressed areas, would be an added financial bonus.

A further benefit of a government-linked equity stake in small high technology firms in depressed regions would be the protection of these firms from the asset-stripping acquisition behaviour of larger firms from outside the region. There is some evidence to indicate that a number of fast-growing firms in the Northern region of England have been acquired in this manner, with a subsequent relocation of the manufacture of their product to another region or abroad (Smith 1982). This is a great loss to a depressed region, both in terms of wasted regional aid of various forms and because of the employment lost or foregone through the product relocation. It is imperative that, in seeking to create from various sources a 'critical mass' of high technology industrial production at a growth pole in a depressed region, there should be a minimum of high technology employment leakage to other areas.

One further advantage of a government-sponsored venture capital agency would be the implicit commitment of such an approach to *medium-term* funding of firms with growth potential. Although product life cycles are short in high technology firms, programmes of research and development on particular projects can easily extend over two or three years. Moreover, the rapid technological advances common in these leading-edge technologies mean that investment finance towards research and development will be a continuing problem that will increase in magnitude as the firm grows. Here grants and loans are of limited value, not only because they are a particularly 'lumpy' form of investment capital flow, but also, and more importantly, because each loan application involves the risk of refusal and, even if successful, does *not* imply any form of long-term commitment to the firm. Clearly, an advantage of the equity stake is that the venture capital agency, as in the case of the private venture capitalist in Silicon Valley, would have a continuing interest in supporting the company.

It is readily apparent that this suggestion for a regional capital investment agency bears many resemblances to the British National Enterprise Board (NEB), now merged with the National Research and Development Corporation (NRDC) to form the British Technology Group (BTG). The efficiency of this constantly changing organisation has been badly affected by the ebb and flow of government policy since its inception in 1975, but it is notable that the organisation has recorded several successes in the funding of high technoloogy small firms, frequently in exchange for equity. While banks always state that no small firm with growth potential is starved of capital, the National Enterprise Board, with its equity option policy and technical back-up service, has been a successful alternative to bank finance for many small firms. Indeed, the prosperity of the now-famous venture capitalists in Silicon Valley, who operate a similar private sector service, is evidence of the inability of local banks to cater for this real investment need in high technology firms, for it is inconceivable that the large profits amassed by venture capitalists in the Bay area would be willingly foregone by bankers. The role of the BTG continues to be revised to conform to the policies of the current British government. But, provided the existing powers of the group are not further diminished, there is clear scope for an adaptation of the regional branches of the BTG to perform the venture capital agency role outlined in this section, *if* they are given greater resources and a large measure of regional autonomy.

Conversely, it might be argued that the evidence of this book has indicated that any form of government involvement would be resisted by small firm entrepreneurs. However, the validity of this view is diminished by the success achieved by both the BTG and the Scottish Development Agency in their funding of small firms. This success demonstrates that government assistance to small firms can be effective when it is delivered by credible organisations with proven business and technical expertise. Entrepreneurs will show greater enthusiasm and respect for the agents of government assistance if these agents can prove their understanding of small firm problems and potentials and do not suffocate the aid process with irrelevant forms and unnecessary delays. The respondents to this survey in both Britain and the United States were rarely adverse to government aid *per se*, but merely abhorred the current over-bureaucratic system of delivery. Under the proposed venture capital agency approach, the exchange of equity for finance, together with the acceptance of a nominated agency director on the company board, will be bearable if the firm owners feel that they are augmenting the overall organisational, technical and financial strength of the firm.

The current environment for small firm finance in Britain is in a state of flux, with large national banks making efforts towards the funding of new high technology firms in a limited manner. Other private venture capital organisations appear on the scene daily in both prosperous and development areas. These new ventures may in time negate the need for the proposed government venture capital agency, or be combined in some way to the advantage of all concerned. However, regardless of the potential growth of private venture capital organisations, the proposed government regional venture capital agency would be a particularly effective means of delivering government aid to small firms in the depressed regions of Britain and the United States, both because of the level of commitment involved in the equity stake approach, and because the tax payer maintains a good chance of recovering his investment. Indeed, the venture capital agency approach would negate much of the national and regional small firm assistance currently available to high technology small firms. The Silicon Valley experience indicates that the 'hands on' exchange of venture capital for equity is particularly conducive to the growth of fast-growing high technology small firms.

Finally, these comments on the requirements of a venture capital agency may be usefully integrated with comments on the encouraging performance of Scotland, a region initially hypothesised as being an archetypal depressed region. Many of the approaches outlined in the preceding pages have been actively pursued in Scotland since the mid-1970s by the Scottish Development Agency (SDA). Beginning from a base of substantial electronic and instrument engineering sectors, the SDA has vigorously pursued a policy of high technology industrial growth, both in terms of attracting inward investment to the region (for example Hewlett and Packard, IBM, Motorola, and Wang), and by promoting the growth of indigenous small firms. It is less surprising, therefore, that the evidence of this book indicates considerable high technology small firm industrial growth in Scotland. Although resource bottlenecks remain a problem, particularly with regard to input materials and services (see Chapter 6) and research and development labour supply (see Chapter 8), Scotland has made considerable progress towards creating, in a depressed region, a high technology industrial base, with considerable prospects of future growth, both from indigenous expansion *and* through inward investment attracted by past success in a circular causative manner.

The activities of the SDA in promoting the growth of small businesses lie behind many of the encouraging signs of growth and expansion noted in the survey. However, the agency would probably admit that its restricted freedom in needing to keep rigidly in line with bank lending policies and bank restrictions on equity holdings in growing firms has inhibited it in the past. In many instances it seems that the private banking sector, unwilling to fill the venture capital gap itself, is none the less antagonistic towards any competing source of finance that might wish to come forward, particularly when that source is based on government funds. However, perhaps the most encouraging aspect of the SDA is its relative power and dynamism when compared with other British and American regional planning authorities. In particular, the agency has power to oversee all industrial development initiatives for Scotland and an evolving commitment to improve its strategies and performance.

Apart from the macro approaches to changing the environment in favour of high technology industry as part of a high technology industrial growth pole strategy, the single most important objective at the level of the firm is to deliver adequate locally based investment capital in a fast and efficient manner. There is evidence from this study to suggest that gaps do exist in the venture capital market for small high technology firms, and that these gaps impede the efficiency of businesses with the greatest growth potential, particularly in the British instance where venture capital is rare. The alleviation of this bottleneck through the methods proposed in this chapter, or similar measures, is the single most important task for government and private agencies concerned with the funding of small firms. However, the SDA example is again useful in introducing a final note of caution to set against the preceding optimistic assertions on the creation of a high technology industrial growth pole. While it is true that the SDA has vigorously pursued a policy designed to diversify the Scottish regional economy through high technology industrial development, a second arm of policy has been concerned with the refurbishment and rationalisation of existing industrial sectors in Scotland. For while the technological improvement of a region will clearly benefit from the development of a high technology industrial growth pole, policy must move forward on a *broad* front in any depressed region to replace *and* repair the older existing industries in decline. Although this book has been specifically concerned with the prospects for small firm industrial growth within high technology sectors in depressed regions, this focus of interest does not deny the need for technological change in the large firms of traditional industries.

Appendix 1: Survey MLH and SIC codes with product categories

British Minimum List Headings

Scientific and industrial instruments and systems: MLH 354

1. Optical instruments. Manufacturing lenses, prisms, and other optically worked elements, telescopes, binoculars, monoculars, microscopes, optical surveying instruments, optical metrological instruments, optical density measuring equipment, opthalmic instruments, photocells and other optical instruments and apparatus. Optical nautical and aeronautical and gunnery control instruments are included, but the grinding of spectacle lenses is classified in Subdivision 2 of MLH 353. Photographic and cinematographic apparatus is classified in MLH 351.

2. Other scientific and industrial instruments and systems. Manufacturing scientific instruments, equipment and systems for sensing, measuring, indicating, recording and/or control of mechanical, electrical (including electronic), and magnetic magnitudes, including simple measuring devices such as pressure gauges, meters. Ultrasonic instruments and equipment are included. Mechanical and electrical medical measuring instruments are included, but other electro-medical equipment is classified in MLH 367. Engineers' gauges are classified in MLH 390.

Radio and electronic components: MLH 364

1. Valves and other active components. Manufacturing electronic valves (including cathode ray tubes), semi-conductors and electronic rectifiers. Glass envelopes are classified in MLH 463.

2. Integrated circuits. Manufacturing thin and thick film passive and hybrid circuits, monolithic semi-conductor circuits. Printed circuits are classified in Subdivision 3 of this heading.

3. Other radio and electronic components. Manufacturing resistors, capacitors, inductors, circuit breakers for electronic equipment, sound reproduction components, printed circuits and other components and assemblies not elsewhere specified. The manufacture of electronic components for equipment classified in MLH 365, Subdivision 2 is included. The manufacture of tape decks (other than those for use with computers and related equipment, which is classified in MLH 366) is included.

Equivalent United States Four-digit Standard Industrial Classification codes chosen for the survey.

SIC 3800 Instruments and related products

3811 Instruments, engineering and scientific; 3822 Environmental controls; 3823 Instruments, process controls; 3824 Fluid meters and counting devices; 3825 Instruments to measure electricity; 3829 Measuring and control devices (NEC); 3832 Optical instruments and lenses.

SIC 3600 Electric and electronic equipment

3612 Transformers; 3622 Controls, industrial; 3629 Electrical industrial apparatus (NEC); 3671 Electronic tubes, receiving; 3672 Cathode ray TV picture tubes; 3673 Electronic tubes, transmitting; 3674 Semiconductors and related devices; 3675 Capacitors, electronic; 3676 Resistors, electronic and electric; 3677 Coils and transformers, electronic; 3678 Connectors, electronic; 3679 Electronic components (NEC).

Appendix 2: The technological complexity variable

This variable was constructed from a more detailed six-level variable that grouped the main output of the survey firms into categories. These categories were ranked according to technological complexity, as follows:

High technology: laboratory instrumentation, process control systems.

Medium technology: single-function measuring and control devices, test equipment manufacture.

Low technology: printed circuit board manufacture, other small electronic components.

Individual survey firms were checked against their allocated level of technological complexity to ensure the categorisation was not misleading. In certain cases firms were reclassified when, for example, a particularly sophisticated test equipment manufacturer's product was considered high technology. However, in the majority of cases the categorisation was deemed valid. While it is accepted that this approach is somewhat subjective, the rankings are based on the author's extensive experience obtained during three separate research studies.

Bibliography

Alonso, W. (1971), 'The Economics of Urban Size', *Papers and Proceedings of the Regional Science Association* , **26**, pp. 67–83.

Anthony, D. (1983), 'Japan', in *The Small Firm: An International Survey*, ed. D.J. Storey, London, Croom Helm, pp. 46–83.

Averitt, R.T. (1968), *The Dual Economy: The Dynamics of American Industry Structure*, New York, Norton.

Bannock, G. (1981), *The Economics of Small Firms: Return from the Wilderness*, Oxford, Basil Blackwell.

Bater, J.H. and Walker, D.F. (1970), 'Further Comments on Industrial Location and Linkage', *Area*, **4**, pp. 59–63.

Birch, D.L. (1979), 'The Job Generation Process', Working Paper, MIT Program on Neighbourhood and Regional Change, Cambridge, Mass.

Bolton, J.E. (1971), *Small Firms: Report of the Commission of Enquiry on Small Firms*, Cmnd 4811, London, HMSO.

Boretsky, M. (1975), 'Trends in US Technology: A Political Economist's view', *American Scientist*, **63**, pp. 70–82.

Boswell, J.C. (1973), *The Rise and Decline of the Small Firm*, London, Allen & Unwin.

Britton, J.N.H. (1969), 'A Geographical Approach to the Examination of Industrial Linkages', *Canadian Geographer*, **13**, pp. 185–98.

Bullock, M. (1983), *Academic Enterprise, Industrial Innovation and the Development of High Technology Financing in the United States*, London, Brand Brothers and Co.

Buswell, R.J. and Lewis, E.W. (1970), 'The Geographic Distribution of Industrial Research Activity', *Regional Studies*, **4** (2), pp. 297–306.

Cameron, G.C. (1979), 'The National Industrial Strategy and Regional Policy', in *Regional Policy*, eds. D. MacLennan and J.B. Parr, Oxford, Martin Robertson, pp. 297–322.

Channon, D.F. (1973), *The Strategy and Structure of British Enterprise*, London, Macmillan.

Chapman, K. (1973), 'Agglomeration and Linkage in the UK Petro-chemical Industry', *Transactions IBG*, **60**, pp. 33–68.

Commission of the European Communities (1980), *Small and Medium Sized Enterprises and the Artisan*, January: reproduced in the Confederation of British Industry, *Smaller Firms in the Economy*, October 1980.

Cooper, A.C. (1970), 'The Palo Alto Experience', *Industrial Research*, May, pp. 58–60.

Cross, M. (1981), *New Firm Formation and Regional Development*, Farnborough, Hants, Gower.

Darwent, D.F. (1969), 'Growth Poles and Growth Centres in Regional Planning — A Review', *Environment and Planning*, **1**, pp. 5–31.

Denison, E.F. (1967), *Why Growth Rates Differ*, Washington DC, Brookings Institute.

Department of Industry Business Statistics Office (1968) (1978), *Business Monitor Report on the Census of Production*, PA 1002, London, HMSO.

Department of Industry Business Statistics Office (1976), *Business Monitor Report on the Census of Production,* PA 1003, London, HMSO.

Deutermann, E.P. (1966), 'Seeding Science-Based Industries', *New England Business Review,* December, pp. 7–15.

Ewers, H-J. and Wettmann, R.W. (1980), 'Innovation-Oriented Regional Policy', *Regional Studies,* **14**, pp. 161–180.

Feller, J. (1975), 'Innovation, Diffusion and Industrial Location', in *Location Dynamics and Manufacturing Activity,* eds. L. Collins and D.F. Walker, London, John Wiley.

Firn, J. (1975), 'External Control and Regional Development: The Case of Scotland', *Environment and Planning* A, **7**, pp. 393–414.

Fothergill, S. and Gudgin, G. (1979), 'The Job Generation Process in Britain', *Centre for Environmental Studies Research Series,* Vol. 32, London.

Fothergill, S. and Gudgin, G. (1982), *Unequal Growth,* London, Heinemann.

Freeman, C. (1974), *The Economics of Industrial Innovation,* Harmondsworth, Penguin.

Fulton, M. and Hoch, L.C. (1959), 'Transport Factors Affecting Location Decisions', *Economic Geography,* **35**, pp. 51–9.

Ganguly, A. (1982), 'Significant Surplus of Births over Deaths', *British Business,* **23** (July), pp. 512–13.

Gibson, J.L. (1970), 'An Analysis of the Location of Instrument Manufacture in the United States', *Annals of the Association of American Geographers,* **60** (2), pp. 352–67.

Gilmour, J.M. (1974), 'External Economies of Scale, Inter-industrial Linkages and Decision Making in Manufacturing', in *Spatial Perspectives on Industrial Organisation and Decision Making,* ed. F.E.I. Hamilton, London, John Wiley, pp. 335–63.

Gitlow, A.L. (1954) 'Wages and the allocation of employment', *Southern Economic Journal,* **21**, pp. 62-83.

Goddard, J.B. (1978), 'The Location of Non-manufacturing Activities within Manufacturing Industries', in *Contemporary Industrialisation,* ed. F.E.I. Hamilton, London, Longman, pp. 62–85.

Goddard, J.B. and Smith, I. (1978), 'Changes in Corporate Control in the British Urban System, 1972–77', *Environment and Planning* A, **10**, pp. 1073–84.

Goodman, J.F.B. and Samuel, P.J. (1966), 'The Motor Industry in a Development Area: A Case Study of the Labour Factor', *British Journal of Industrial Relations,* **9**, pp. 335–65.

Gudgin, G. (1978), *Industrial Location Processes and Regional Employment Growth,* Farnborough, Hants, Saxon House.

Hall, P. (1962), *The Industries of London,* London, Hutchinson.

Johnson, P.S. and Cathcart, D.G. (1979), 'New Manufacturing Firms and Regional Development: Some Evidence from the Northern Region', *Regional Studies,* **13**, pp. 269–80.

Karaska, G.T. (1969), 'Manufacturing Linkages in the Philadelphia Economy: Some Evidence of External Agglomeration Forces', *Geographical Analysis,* **1** (4), pp. 354–69.

Keeble, D. (1976), *Industrial Location and Planning in the United Kingdom,* London, Methuen.

Keeble, D. (1978), 'Industrial Decline in the Inner City and Conurbation', *Transactions IBG,* **3** (1), pp. 101–14.

Kondratiev, N.D. (1935), 'The Long Waves in Economic Life', *Review of Economic Statistics,* **17** (November), pp. 105–15.

Lever, W.F. (1974), 'Manufacturing Linkages and the Search for Suppliers and Markets', in *Spatial Perspectives on Industrial Organisation and Decision Making,* ed. F.E.I. Hamilton, London, John Wiley, pp. 309–33.

Little, A.D. (1977), 'New Technology-Based Firms in the United Kingdom and the Federal Republic of Germany', Report prepared for the Anglo-German Foundation for the Study of Industrial Society, London.

Lloyd, P. and Dicken, P. (1977), *Location in Space*, New York, Harper Row.

Lösch, A. (1954), *The Economics of Location*, New Haven, Yale University Press.

Luttrell, W.F. (1962), *Factory Location and Industrial Movement*, London, NIESR.

Macmillan Committee (1931), *Report of the Committee on Finance and Industry*, Cmnd 3897, London, HMSO.

Martin, J.E. (1966), *Greater London: An Industrial Geography*, London, Bell.

Mason, C. (1983), 'Some Definitional Difficulties in New Firm Research', *Area*, 15 (1), pp. 53–60.

McDermott, P. (1976), 'Ownership, Organisation and Regional Dependence in the Scottish Electronics Industry, *Regional Studies*, 10, pp. 319–35.

Meeks, G. (1977), *Disappointing Marriage: A Study of the Gains from Merger*, Cambridge University Department of Applied Economics Occasional Paper No. 51, Cambridge University Press.

Monopolies and Mergers Commission Report (1982) *The Hong Kong and Shanghai Banking Corporation Standard Chartered Bank Limited, The Royal Bank of Scotland Group, A Report of the Proposed Merger*, Cmnd 8472, London, HMSO.

Morse, R.S. (1976), *The Role of New Technical Enterprises in the US Economy*, Report of the Commerce Technical Advisory Board to the Secretary of Commerce, January.

Moseley, M.J. and Townroe, P.M. (1973), 'Linkage Adjustment Following Industrial Movement', *Tijdschrift voor Economische en Sociale Geografie*, 64 (3), pp. 137–44.

Nicholls, V. (1969), 'Growth Poles: An Evaluation of Their Propulsive Effect', *Environment and Planning*, 1, pp. 193–208.

NRST (1976), *Strategic Plan for the Northern Region*, Vols. 1–5, Newcastle upon Tyne, Northern Region Strategy Team.

Oakey, R.P. (1979a), *An Analysis of the Spatial Distribution of Significant British Industrial Innovations*, Discussion Paper No. 25, Centre for Urban and Regional Development Studies, University of Newcastle upon Tyne.

Oakey, R.P. (1979b), 'Labour and the Location of Mobile Industry: Observations from the Instruments Industry', *Environment and Planning* A, 11, pp. 1231–40.

Oakey, R.P. (1979c), 'The Effect of Technical Contacts with Local Research Establishments on the Location of the British Instruments Industry', *Area*, 11, pp. 146–50.

Oakey, R.P. (1981), *High Technology Industry and Industrial Location*, Farnborough, Hants, Gower.

Oakey, R.P. (1983), 'New Technology, Government Policy and Regional Manufacturing Employment', *Area*, 15 (1), pp. 61–5.

Oakey, R.P. and Seaman, M. (1981), *The Location of New and Newly Relocated Industrial Firms in the Borough of Ealing*, Oxford Polytechnic Discussion Paper in Geography No. 16.

Oakey, R.P., Thwaites, A.T. and Nash, P.A. (1980), 'The Regional Distribution of Innovative Manufacturing Establishments in Britain', *Regional Studies*, 14, pp. 235–53.

Oakey, R.P., Thwaites, A.T. and Nash, P.A. (1982), 'Technological Change and Regional Development: Some Evidence on Regional Variations in Product and Process Innovation', *Environment and Planning* A, 14, pp. 1073–86.

Parsons, G.F. (1972), 'The Giant Manufacturing Corporations and Balanced Regional growth in Britain', *Area*, 4 (2), pp. 99–103.

Pirroux,F. (1955), 'Note sur la notion de pôle de croissance', *Economie Appliqué*, **8** (1 and 2).

Prais, S. (1976), *The Evolution of Giant Firms in Britain,* Cambridge, Cambridge University Press.

Pred, A.R. (1965), 'The Concentration of High Value Added Manufacturing', *Economic Geography*, **41**, pp. 108–32.

Pred, A.R. (1967), *Behaviour and Location,* Lund Studies in Geography, 2 vols, Lund, Lund University.

Rabey, G. (1977), Contraction Poles: *An Exploratory Study of Transitional Industry Decline Within a Regional Industrial Complex,* Discussion Paper No. 3, Centre for Urban and Regional Development Studies, University of Newcastle upon Tyne.

Riley, R.C. (1973), *Industrial Geography,* London, Chatto and Windus.

Roberts, E.B. (1977), 'Generating Effective Corporate Innovation', *Technology Review,* **80** (1), pp. 27–33.

Robertson, A. (1974), 'Innovation Management', *Management Decisions,* **12**, pp. 329–73.

Rothwell, R. (1982), 'The Role of Technology in Industrial Change: Implications for Regional Policy', *Regional Studies,* **16** (5), pp. 361–9.

Rothwell, R. and Zegveld, W. (1982), *Innovation and Small and Medium Sized Firms,* London, Frances Pinter.

Sant, M.E. (1975), *Industrial Movement and Regional Development,* Urban and Regional Planning Series, Vol. II, London, Pergamon Press.

Sappho Project (1972), *Success and Failure in Industrial Innovation,* Science Policy Research Unit, Centre for the Study of Innovation, London.

Schmookler, J. (1972), *Patents Invention and Economic Change,* Cambridge, Mass., Harvard University Press.

Segal, M. (1962), *Wages in the Metropolis,* Cambridge, Mass., Harvard University Press.

Senker, P. (1979), *Skilled Manpower in Small Engineering Firms: A Study of UK Precision Press Tool Manufacturers,* Science Policy Research Unit, Sussex University.

Senker, P. (1981), 'Technical Change, Employment and International Competition', *Futures,* June, pp. 159–70.

Shapero, A. (1980), 'The Entrepreneur, the Small Firm and Possible Policies', Paper presented at Six-Counties Programme Workshop on Entrepreneurship, Limerick, Ireland (PO Box 215 TNO, Delft, The Netherlands).

Smith, D.M. (1970), 'On Throwing Weber Out With the Bathwater; A Note On Industrial Location and Linkage', *Area,* **1**, pp. 15–18.

Smith, I.J. (1979), 'The Effect of External Takeovers on Manufacturing Employment Change in the Northern Region Between 1963 and 1973', *Regional Studies,* **13**, pp. 421–37.

Smith, I.J. (1982), 'Some Implications of Inward Investment Through Takeover Activity, *Northern Economic Review,* **2**, pp. 1–5.

Solow, R.M. (1957), 'Technical Change and the Aggregate Production Function', *Review of Economic Statistics,* **39**, pp. 312–20.

Storey, D.J. (1982), *Entrepreneurship and the Small Firm,* London, Croom Helm.

Storey, D.J. (1983), 'Small Firm Policies: A Critique', *Journal of General Management,* **8** (4), pp. 5–19.

Taylor, M. (1970), 'Location Decisions of Small Firms', *Area,* **2**, pp. 51–4.

Taylor, M. (1973), 'Local Linkage, External Economies and the Ironfoundry Industry of the West Midlands and East Lancashire Conurbations', *Regional Studies*, 7, pp. 387–400.

Thomas, M.D. (1975), 'Growth Pole Theory, Technological Change and Regional Economic Growth', *Papers of the Regional Science Association*, 34, pp. 3–25.

Thompson, J.H. (1983), 'The United States of America', in *The Small Firm: An International Survey*, ed. D.J. Storey, London, Croom Helm, pp. 7–45.

Thwaites, A.T. (1978), 'Technological Change, Mobile Plants and Regional Development., *Regional Studies*, 12, pp. 445–61.

Thwaites, A.T., Oakey, R.P. and Nash, P.A. (1981), *Technological Change and Regional Development in Britain*, Final Report, Centre for Urban and Regional Development Studies, University of Newcastle upon Tyne.

Townroe, P. (1971), *Industrial Location Decisions*, Occasional Paper No. 15, Centre for Urban and Regional Studies, University of Birmingham.

Townroe, P.M. (1975), 'The Labour Factor in the Post-move Experiences of Mobile Companies, *Regional Studies*, 9, pp. 335–47.

Townsend, J., Henwood, F., Thomas, G., Pavitt, K. and Wyatt, S. (1981), *Innovations in Britain Since 1945*, Occasional Paper Series No., 16, Science Policy Research Unit, Sussex University.

United States Bureau of Census (1968) (1978), *County Business Patterns—California.*

Vernon, R. (1966), 'International Investment and International Trade in the Product Cycle', *Quarterly Journal of Economics*, 80, pp. 190–207.

Von Hippel, E. (1977), 'Successful and Failed Co-operative Ventures: An Empirical Analysis', *Industrial Marketing Management*, 6, pp. 163–74.

Weber, A. (1909), *Theory of the Location of Industries*, Chicago, University of Chicago Press.

Wilson Committee (1979), *The Financing of Small Firms*, Interim Report of the Committee to review the functioning of the financial institutions, Cmnd 7503, London, HMSO.

Wise, M.J. (1949), 'On the Evolution of the Gun and Jewellery Quarters in Birmingham', *Transactions IBG*, 15, pp. 57–72.

Wood, P. (1969), 'Industrial Location and Linkage', *Area*, 2, pp. 32–9.

Zegveld, W. and Prakke, F. (1978), 'Government Policies and Factors Influencing the Innovative Capability of Small and Medium Enterprises', Paper presented to the Committee for Scientific and Technological Policy, OECD, Paris.

Index

Agglomeration
  and high technology industries, 48, 50-2,
      52-3, 75, 87-91, 141, 146-52
  principles of, 42-6
Alonso, W., 51, 123, 147, 167
Anthony, D., 6, 167
Averitt, R.T., 16, 167

Banks
  and investment finance 128, 135, 137-9
  and new firms, 129, 130
Bannock, G., 7, 8, 167
Bater, J.H., 45, 167
Belgium
  employment in small manufacturing
    firms, 5
Birch, D.L., 4, 167
  Birch report, 4
Bolton, J.E., 4, 8, 107, 129, 167
  Bolton Committee, 4
Boretsky, M., 12, 167
Boswell, J.C., 30, 167
Brazil, 28
Brittan, J.N.H., 45, 167
Britain, 55-6
  declining industries in, 28
  electronics industry, 47
  high technology industry, 58, 60
  innovativeness in manufacturing
    industry, 17-22
  instruments industry, 74
    shopfloor labour in instruments
    industry, 109
  investment finance for new small
    businesses, 129
  motor-vehicle industry, 47
  small manufacturing firms, 9-12
  see also British government, British
    National Enterprise Board,
    British Technology Group
British government, 55

aid to high technology industry in
  Scotland, 67
assistance to declining industries, 28
incentives for relocation, 44
  electrical and mechanical engin-
    eering, 48
small firm assistance, 4
British National Enterprise Board (NEB),
  159
British Technology Group (BTG), 159
Bullock, M., 37, 56, 106, 123, 139, 150, 167
Businessmen
  skills, 30-2
  see also Entrepreneurs
Buswell, R.J., 48, 55, 167

Cameron, G.C., 9, 167
Canada
  employment in small manufacturing
    firms, 8
Cathcart, D.G., 9, 33, 37, 128, 168
Channon, D.F., 7, 16, 165
Chapman, K., 151, 167
Commission of the European Com-
  munities, 5, 167
Computer numerically controlled machines
  (CNC), 14
Consultants, see External consultants
Cooper, A.C., 33, 37, 104, 105, 150, 167
Cross, M., 130, 167
Customer relationships, 81, 82, 86-7

Darwent, D.F., 151, 167
Denison, E.F., 12, 167
Denmark
  employment in small manufacturing
    firms, 5
Department of Industry (Britain), 5, 10,
  165, 168
  Offices and Service Industries Scheme
    (OSIS), 155

Rabey, G., 35, 170
'Raw materials' in high technology industries, 75, 76
Recruitment, 117, 119
Regional environment, 55
Regional industrial employment
contribution of survey industries to, 58-65
in San Francisco Bay area, 64, 65; in Scotland, 60, 62; in South East (England), 62
Regional investment
decline in, 26
Regional small firm growth, 12-17
effect of industrial structure on, 24-41
Regional small firm innovation
and high technology agglomerations, 52-3
high technology industries and local physical environment, 68
high technology industries and local sourcing, 79
measures of regional innovativeness in survey regions, 68-71
Regional variations, 12-13
in Britain, 9-12
in British manufacturing industry, 18
in proportion of shopfloor workers in survey firms, 112-14
in small firm product innovation levels, 17-22
Relocation, 44-5, 47, 48, 50-1, 51-2
Research and development
and firms with high technology product base, 144
and high technology industrial agglomerations, 149-50
and innovation, 92-107, 117, 125
external technical information, 104-6; internal research and development, 96-104; necessity for, 106-7; research and development cycles, 92-6
and 'long product life cycle' based firms, 143
and subcontract firms, 142
government-funding policies, 156
investment in, 128, 142
problems over cost of, 126, 127

recruitment of personnel, 120
Riley, R.C., 43, 168
Risk, 93, 95, 125, 126, 127, 130
Roberts, E.B., 7, 30, 107, 170
Robertson, A., 76, 168
Rothwell, R., 8, 12, 17, 21, 29, 39, 107, 170
Route 128, 51, 56, 151

Samuel, P.J., 47, 168
San Francisco Bay area of California, 56
high technology industries, 58, 59, 63-5
agglomeration advantage, 146, 148; customer feedback, 86-7; external acquisition, 119, 120, 121; external technical information, 104, 105, 107; finance innovation and the new firm, 129, 130, 132; in-house training, 118; innovativeness, 68, 69, 70; internal research and development, 96, 98, 99, 100, 101, 102, 103; labour shortages, 115, 116, 117, 147; labour shortage and innovation, 123; local physical environment, 67, 68; local sourcing and innovation, 77, 78, 79, 80, 91; output linkages and innovation, 81, 82, 83, 84, 85, 86; proportion of shopfloor workers, 113, 114; sources of investment finance, 133, 135, 136, 137, 138, 139, 140, 142, 147
product-based small firms and employment, 17
Sant, M.E., 29, 45, 170
Sappho Project, 12, 170
Schmookler, J., 12, 170
Scotland, 3, 24, 55
employment in small manufacturing firms, 10, 11
high technology industries, 58, 59-62, 161
customer feedback, 86-7; external acquisition, 119, 120; external technical information, 105; finance innovation and the new firm, 129, 130, 132; in-house training, 118; innovativeness, 68, 69, 70, 71; internal research and development, 96, 97, 98, 99, 100, 101, 102,